THE WAR ON THE
CIVIL & MILITARY FRONTS

The Lees Knowles Lectures on Military History for 1942

THE WAR ON THE CIVIL AND MILITARY FRONTS

by

MAJOR-GENERAL G. M. LINDSAY
C.B., C.M.G., D.S.O.
Colonel Commandant, Royal Tank Regiment:
Deputy Regional Commissioner,
South-Western Region

CAMBRIDGE
AT THE UNIVERSITY PRESS
1942

CAMBRIDGE
UNIVERSITY PRESS

University Printing House, Cambridge CB2 8BS, United Kingdom

Published in the United States of America by Cambridge University Press, New York

Cambridge University Press is part of the University of Cambridge.

It furthers the University's mission by disseminating knowledge in the pursuit of education, learning and research at the highest international levels of excellence.

www.cambridge.org
Information on this title: www.cambridge.org/9781107418844

First published 1942
First paperback edition 2014

A catalogue record for this publication is available from the British Library

ISBN 978-1-107-41884-4 Paperback

Contents

Preface

These lectures were written between 1 December 1941 and 1 February 1942 in the period marked by the opening of the Russian counter-offensive, General Rommel's retirement and second come-back in Libya and the overrunning of Malaya by Japan. The canvas of the war is world-wide, the kaleidoscope dazzling. The course of our fortunes since September 1939 has been chequered, threat and recovery, counter-offensive and set-back alternating with bewildering speed. At one and the same time battles are in progress in Arctic waters, snow-bound steppes, sand-swept desert and steaming jungle. Whether history will one day say that here political strategy committed us to battle at a tactical disadvantage or that there tactical and technical deficiencies brought political strategy to grief, it is yet too soon to conjecture. For one thing the facts are not available. All that we can do now is to try and bring the scene into something like perspective and to study it through the glasses which it is the endeavour of military science to supply—with glasses which take a wide sweep from the munition plant to the battlefield. But we may say with confidence that never before has our country been confronted with problems of such magnitude. By the time-table of the late war we are now on the eve of Vimy Ridge, which was the beginning of the penultimate stage. In 1939 we had no Expeditionary Force comparable in readiness or modernity to that of 1914. In 1940 we lost the continental base which in 1915 we held till we advanced from it to final victory. In 1941 in the Middle East we were dislodged from the equivalent of Salonika. Once and only once have we been able to fight from a logical base, namely in the air battle of 1940 over Britain. Our main military effort has been made perforce from Suez under conditions markedly less favourable than those of 1914–18; for then we shared the Mediterranean with France and Italy, but now, while the enemy operates on inner lines with only a short sea passage from Sicily, we have to reinforce our army by a sea route of some thousands of miles. A temporary retirement may bring the enemy back to fresh stores of heavier weapons and thus in the end be a gain to him, whereas a temporary retirement may be imposed on us for the reason that the Far East can be reinforced more quickly from Suez than from Britain or America. That we have maintained continuous action on land, taking blow after blow from so many quarters, has been possible only for two reasons: the service of the Navy in convoy and interception, and the service of the Air Force in arsenal defence and long-range offensive. Meanwhile the land forces of Britain and the Empire, facing difficulties both material and psychological—the material difficulties of late rearmament and the psychological difficulties of distance from home and familiar terrain—have never failed to give of their utmost, even when that utmost involved replenishing an ally to the prejudice of themselves. Once again it is too soon to pronounce judgment even on a single theatre of the war, but assuredly when judgment is passed, it will pay tribute to the resilience and ubiquity of our military effort.

When a country's experience is drawn from so many theatres, the task of extracting the right lessons from it is enhanced both in difficulty and essential importance. Therefore we must try to find mental life-lines and to grasp the central truths of a strongly centrifugal scene. History can be either a hindrance or a boon. It is a hindrance if we discourage ourselves by dwelling on the easier victories of the past or by fretting over recent disappointments. It is a boon if we probe it so deeply that it reveals enduring lessons and furnishes principles on which to think and act. Making history has two very difference senses. The hero, or assassin, makes history in one sense. The stream of things, with its heroisms and crimes among them, makes history in another and deeper sense. The climber's rope has running through it a red strand, signifying that it has been tested to hold. Can we in the study of history find a military rope that will stand the present strain? It is worth the search; for the rope that will hold the military arm will hold the civil arm also—will hold, in fact, the whole people. The University of Newton and Pitt, of Castlereagh and Darwin will concede at least that alike from the angles of science and statecraft military history has work to do which is very proper to a principal place of Higher Education and Research.

When I had the honour, and indeed I knew full well what a great honour it was, to be invited to deliver the Lees Knowles lectures, it was only natural that I should wonder if I had the background necessary for so onerous a task. I accepted because I felt that, given the ability to translate it into words, I had the experience, perhaps almost unique in its way, which entitled me to deal with modern war both on the civil and military fronts.

It will be generally admitted that the two outstanding weapons of modern land warfare are the machine gun and the tank. The first I had made my especial study during the last war, during which, amongst others, I held the appointments of Staff Officer to the Machine Gun Training Centre at Grantham, Chief Instructor to the G.H.Q. Machine Gun School in France, and Army Machine Gun Officer to the First Army during the last year of the war. In addition I had written the official manual on the Tactical Employment of Machine Guns, S.S. 192 of January 1918. I had therefore had great opportunity for studying the evolution of the weapon that was the undisputed master of the battlefield from the first day of the war until the development of that other master weapon, the tank, deprived it of pride of place.

From the end of the war onwards I devoted myself to the study of Mechanized Warfare or, as we later always called it, Armoured War. In this sphere again I had an almost unique opportunity for study, for I was successively Commander of the First Armoured Car Group in Iraq, Chief Instructor Royal Tank Corps Central Schools, Inspector Royal Tank Corps at the War Office, Brigadier-General Staff Egypt with a special mandate from the then C.I.G.S. to study mechanized warfare in the desert, and Commander of the 7th (Mechanized Experimental) Infantry Brigade on Salisbury Plain. I was then promoted Major-General and, while on half-pay, was commissioned by the War Office to set the paper for the Staff College examination which deals with Modern Warfare and to write a new manual on Armoured and Mechanized War.

From this I proceeded to India as Commander of the Presidency and Assam District, which comprises the provinces of Assam, Bengal, Bihar and Orissa. Here I dealt with the Governors, Civil Service and Police of these four provinces as well as with a considerable force of soldiers, regular and irregular, British and Indian. In addition, with the assistance of Sir John Anderson, then Governor of Bengal, and later with the collaboration of his successor the late Lord Brabourne, I had much to do with organizing the Youth Movement in Bengal. These experiences brought me into close contact with every class, official and unofficial, civil and military, British and Indian.

From them all, throughout my time in India, I received nothing but help, kindness and a collaboration leading to many new friendships, Indian and British, which made my stay in India one of the happiest times of my life, and gave me an increased knowledge of humanity which has stood me in good stead. For the first six months of this war I commanded a Highland Division and in April 1940 I was given my present appointment. For the first time I came into intimate contact with the local government of England and its officials: and my experience here, like that in India, has been one of ever-increasing kindness, help and collaboration, with an ever-growing circle of new friendships. It has indeed been a time during which I have learnt much that was new to me, and done much hard work, but in an atmosphere that has made the learning of the deepest interest and the work itself a real pleasure. In my present duties I have had the pleasure and the advantage of having as my Regional Commissioner my old friend Sir Hugh Elles of Tank Corps fame in the last war. Therefore, when I came to prepare these lectures, I could truly feel that I had the background and experience for dealing with both the civil and military fronts of modern war.

My time at Cambridge was indeed a pleasant one; for, as the guest of Professor G. M. Trevelyan, the Master of Trinity, and of Sir Will Spens, Master of Corpus Christi and Regional Commissioner, I received all the kindness and help it was possible for them to extend to me. But the man to whom, above all, I owe the deepest debt of gratitude is my old friend of machine gun days in the last war, Mr C. R. Fay of Christ's. If these lectures have achieved any merit, it is due to his unfailing help, collaboration and inspiration. Without his help I could have achieved little; as a result of it I trust that what I have now written will be of some value to those who may think it worthy of their consideration.

GEORGE LINDSAY

BRISTOL
March 1942

Chapter I

FEATURES AND FUNDAMENTALS

Part I. THE FEATURES OF MODERN WARFARE

The principles of war are eternal, but the factors with which they have to deal are undergoing an incessant evolution. von der Goltz.

1. Introduction

The features (or factors as von der Goltz calls them) of modern warfare are the effect of certain causes, which can only be fully understood when they are put in a historical setting. War, indeed, is destructive in its purpose and wasteful in its results, and therefore as an ideal utterly irrational.[1] None the less, as an activity it is severely rational, and if we do not understand its evolution, we shall be incapable of appreciating the significance of its present features and the way in which the 'eternal principles' have to be applied to new conditions. The principles of war concern certain fundamentals, which apply to all war, ancient, medieval and modern. But they have to be adapted to the features of the warfare with which they are concerned. The features are the conditioning environment, the principles the central core. In final analysis these principles are the rationalized response of the human mind to an environment which is under constant change as the result of inventions and new technique. From this it follows that the principles of war, however immutable, must constantly be re-examined

[1] War, however, acts as a mental stimulus and draws into its machine large numbers of scientists, inventors and others who, in normal times, are engaged in peaceful activities. These, with the financial and other assistance which war places at their disposal, produce technical improvements and material output at a pace unknown in peace. The increased tempo of war is of value afterwards, the great advance in mechanical transport both in the air and on land after the last war being a case in point. Furthermore, in spite of the waste of resources which war involves, it does stimulate the desire for, and perhaps even the progress of, social reform. It has been a factor in the great advance in social services of all kinds, with the resulting betterment in the housing, health and living conditions of the people. But, of course, these results cannot be expected to ensue, if war among the great nations becomes endemic.

and re-applied to the new environment. A man might know exactly the geography and sequence of a dozen major campaigns, but unless he grasps the big ideas running through all wars, he will flounder in detail and be no wiser for his knowledge. Indeed, it may be a positive handicap to him; for it may cause him to rely on weapons and devices which once were serviceable, but now are out of date. This is a particular danger for the peace-loving democracies; because when they have emerged from a war, they hate the memory of it so greatly that they put it from their minds as a thing they have finished with for good. Then, when they are dragged into another war, they are called upon to put on old garments which are moth-eaten, or new garments which as yet do not fit. So much is this the case that one might almost add to the features of modern warfare the probability that, though in the end the democracies may win, at the outset they will lose ground. It was so in the Great War of 1914–18, it is so in the Greater War of to-day. It is a method which is terribly expensive in human life and treasure, however innocent and in a sense praiseworthy the original motive may have been. But it is no use crying over spilt milk. I mention it only to make you realize the magnitude of the task before us.

You, the members of a great University, whether you are in uniform or plain clothes, must be a great thinking factory for the fashioning of a new mental armour. For to re-study and re-think is as important as to re-arm. In peacetime you have your two great divisions of study, each with its own technique—the humanities of literature, history and philosophy: the sciences of mathematics, physics, chemistry and the like. These in wartime should become what I may term the tactical and technical sides of your war effort.[1] Both sides are of equal importance. Indeed, the efficacy of each depends on the one keeping step with the other. It is wrong to suppose that during a war there is no time for historical thinking. We can decide neither the organization nor the tactical role of any machine unless we place it in its historical setting: we cannot arrive at the correct tactical or technical application of the machine unless we understand its evolution, or if it is a new machine, the evolution

[1] Cf. Peter Kapitza, F.R.S., in *Soviet War News* (3 March 1942): 'Soviet botanists are devising camouflage corresponding to seasonal changes in flora. Soviet historians are helping to fight the brazen pseudo-scientific Fascist propaganda.'

of the *idea* which gave it birth. Thus much for the relation of features (or factors) to principles and the need for a historical approach.

2. The Main Features

(*a*) Modern War is World-wide

One war kindles another till the whole world is ablaze. Every part of the world is now in close communication, linked together by wireless, cable, aeroplanes, railways, steamships and motor transport. The world is interdependent as never before, and once major war starts it draws into its vortex one country after another until only a few quasi-independent neutrals are left. War on this scale involves high strategy: strategy involves geography. Maps are of two kinds, physical and political; and a political map is the product of history. Why is the North Atlantic the lifeline of England? Why are Egypt and the Suez Canal the key points of our operations in the East? The reason is that we are an island empire and that these two theatres are on the base line of that empire, along which it is fed and reinforced. As an island without an empire, we could make no resistance in depth, for we have not the space in which to do it. That is why command of the sea has been so important for us and why the Navy has been the senior arm. This base line existed in embryo in Napoleonic times. The West Indian waters then were what the North Atlantic is now. The route to the East was around the Cape, but the importance of the Mediterranean was realized; and in this sea at either end, at Trafalgar and at Aboukir Bay, Nelson won for us the victories which gave us command of the High Seas. When the Suez Canal was opened in 1869, the base line ran direct down the Red Sea to India and beyond.

In the War of 1914–18, with the German fleet in being and Turkey on the enemy's side, the Battle of Jutland and Allenby's victories in Palestine kept the whole base line open. In the present war the Atlantic patrol and the Libyan campaign are the counterpart of Jutland and Palestine, of Trafalgar and Aboukir Bay. The new feature is the emergence of the Pacific as a major theatre of war, consequent upon Japan's declaration of war upon us and the United States. This caused major war to span the world. What the Suez and the Indian Ocean are to us, the Panama and the Pacific are to the United States. In 1939 none foresaw the quick and complete collapse of our principal ally, France. This by its exposure of the Mediterranean route has gravely impeded our operations in the Middle East, compelling us to use the old round-about route alike for munitions and troops. A hostile or severely neutral South Africa would have made this barely possible, but South Africa played her part nobly and enabled us to link our effort with that of herself, of India, of Australia and New Zealand.

The immediate implication of world-wide war to-day is twofold. First, Britain with Canada in her rear, and the Middle East with India and Australasia in its rear, must depend to an increasing extent upon themselves for munitions as well as for men. Secondly, America, being in the War, is now at one with us in a sense that no neutrality, however benevolent, could have permitted. Her forces in Britain and Australia emphasize this fact.

The present situation throws into even greater relief the historic importance of the Suez area—Egypt, Sinai, Palestine. Its importance to anyone who essays to hold empire east and west of this area has been considered paramount from the days of the Pharaohs and the Great Kings onwards. That is why Palestine has been called the cockpit of Asia and why the route followed by Allenby's Cavalry in their advance through the Musmus Pass has been described as one by which from time immemorial 'the caravans of peace and the cohorts of war' have moved. To-day the Suez area is a nodal point of communications, a centre of supply of oil and other war material, a bastion between Europe and our Eastern Empire. Lose it and the Axis fleets join hands in the Red Sea, India is isolated, and contact with Australasia is severed. Thus its historical importance is confirmed and the necessity for its retention is emphasized.

(*b*) Modern Warfare is Mechanical

The Napoleonic War was the first great war to be fought within the era of modern industrialism, but the technique of industrialism was then too new to affect significantly the course of it. Ordnance, indeed, was improved, monetary outlay, consequent on greater commerce, was on a scale hitherto unknown, but at sea it was still a war of wooden ships and sail, and on land it was, as of old, a warfare of manœuvre with cavalry and infantry as the dominant arms. The most significant technical development was the specialization of artillery into a separate arm, employed in mass, with tactics appropriate to such a conception; and it was Napoleon who showed the way here.

By the Great War of 1914–18 the position was greatly changed. The Navy was fully mechanized. The Army, which began the war with only a small contingent of mechanical transport, gradually built up a vast amount of such transport in addition to the true mechanization represented by tanks, armoured cars, motor machine-gun units, etc.[1] The Cavalry, indeed, used their mobility with success in Palestine and to a small extent elsewhere, but on the main Western Front the entanglement of barbed wire, trenches and mud, together with their vulnerability to machine-gun fire, rendered them practically ineffective in their normal role. The Infantry, therefore, had to bear the brunt of the effort in attack as well as in defence. New tactics were worked out by the Artillery and the Machine Guns (which had now emerged as a separate arm intermediate between the Artillery and Infantry) to assist the Infantry in their task of breaking through the trench system. Thus was developed the preliminary bombardment and the creeping barrage, in which the Machine Guns shared. New eyes were given to the Infantry by the use for the first time in warfare of the aeroplane, which in four years developed from a problematical novelty to a reliable and indispensable agent of war. Communications by telephone and wireless underwent a similar mechanical revolution.

The practical consequences of all this new mechanism were profound and may be summarized thus:

(i) The Infantry took on new tasks. To their normal equipment of rifle and bayonet they added the hand and rifle grenade, the light machine gun, the mortar, and at times the dagger, club and pistol of trench raiding parties—parties which were the forerunners of the Commandos and Guerrillas of to-day.

(ii) The work of the several arms in conjunction required increasingly elaborate organization. Hence the set-piece attack, the rehearsal of such attacks, the emphasis on synchronization.

(iii) The supply of an army or armies attacking on a wide front became an increasingly heavy task. Artillery and Machine Guns putting up continuous screens of fire expended an unprecedented amount of ammunition.

[1] For detail of the mechanised component see pp. 19, 28–9, 49, 52.

The task of delivering supplies of all kinds to the fighting troops absorbed transport facilities to an ever greater extent. On wet ground the effect of a prolonged bombardment was to convert it into a sea of mud, which defeated, or at any rate conflicted with, the object of the bombardment, namely to permit of an advance over and beyond the trench system. The result was an impasse, which apparently was a genuine impasse: and the consequence a war of attrition in which at enormous cost in human life the man-power of one side exhausted the man-power of the other. But the impasse was only apparent. For there was a solution; namely the tank, a British invention, a moving fortress of fire-power, which could not be mown down, as waves of infantry were, by machine-gun fire. On this, which has been my study for the last twenty years, I shall have much to say later.

(*c*) Modern Warfare is Totalitarian

The meaning of totalitarian war is not simply war as conducted by the totalitarian powers, but war in which the whole effort of the country, its human as well as its material resources, its civilian as well as its armed services are simultaneously and fully engaged. It is thus war on two fronts, military and civil; and the totality of the effort is well expressed in an extract from a recent bulletin.

> 'I liken the war effort to a triangle, the apex of which is the fighting forces, the base of which represents the ordinary population of the country. If the base of the triangle should disintegrate, the apex will fall to the ground.'

When Germany, following standard routine, suddenly declared war without provocation on Russia, there were those who thought that the episodes of the West in 1940 would be repeated in the East in 1941; or at any rate that even if Russia did fight hard, she would be finished in a couple of months. Now, no one will deny that from a military point of view Germany gained great initial victories over Russia, but equally no one will deny that she failed to gain a decision. The triangle still holds, the apex as well as the base; and the sources of Russia's stalwart stand were these.

(i) She had the breadth of territory in which to resist with all the arms at her disposal, and on which to offer a cushion of resistance in depth. In addition she had the almost unlimited man-power which enabled her to plan her defence in the greatest depth and at the same time to supply the troops required for those constant counter-attacks, on scale both large and small, which are essential features of the defence against the German armoured offensive. She was therefore never forced into the luckless position in which we found ourselves, first at Dunkirk and then at Crete.

(ii) She had unity within herself as the result of her drastic revolution. She was purged of her fifth columnists, i.e. of the people who in Western Europe were concealed in the base of the triangle and by their treachery brought the apex down.

(iii) She had already adopted, almost as a God, the machine spirit. She had mechanized and integrated her industries and agriculture. In Tsarist days she had given little evidence of this aptitude, and in making the change she imported the technique and technicians of Germany, England and America. But she carried it through to the logical end, so that now she is able not only to replace her lost industrial areas by organized evacuation at high speed to points further east, but also to take over and make full use of tanks and planes delivered from Britain and America.

Thus equipped materially and morally Russia carried out with success the two most difficult operations of war. First, she conducted a retreat over hundreds of miles in face of many heavy defeats and yet maintained her army intact with its morale high. Secondly, she launched her big counter-offensive just at the moment when the enemy, with his communications stretched almost to breaking point, had launched his all in a last unsuccessful effort. Thereby she forced the Germans into an operation not of their own design, namely to disengage and to turn from the strategic offensive to the defensive on a huge front and under climatic conditions for which it appears that they were not prepared.

The position of Britain in totalitarian war is that, just as in 1914–18, all her resources are being devoted to the war by land, air and sea, and these resources, in association with those of the Empire and the U.S.A., are being employed on a series of foreign fronts. In addition, in the present war, hostilities are closer to our midst, even though for the time being our overseas fighting is much farther away. In this war Britain became the headquarters of the anti-Nazi front, and incidentally to this the fortress aerodrome, the main base for air and sea attack on Germany. The preservation of this island is not simply a matter of self-defence for Britain. It is the fundamental condition of ultimate victory for all freedom-loving peoples. The defence is both civil and military, and the two must be co-ordinated in the total defence if we are to make ourselves secure against invasion, air-borne or sea-borne, and aerial bombardment. The civil effort is conducted under conditions that approximate to those of a campaign. In the late war, apart from spasmodic Zeppelin and air raids, England was outside the war zone, except for the rumble which at times reached her of the incessant gunfire in France. Now she is in it, or rather under it. For the sky is filled at all hours of the day and night with our aeroplanes in operation or in training. The Battle of Britain was fought over our heads. Many of our great towns have been heavily bombed and severely damaged; and though we have not yet been invaded, we have to be on the constant alert against it. If invasion comes, the serious task awaits us of re-arranging ourselves within our narrow island limits in such a way as to continue production and fight the enemy at one and the same time. The black-out, night bombing, fire-watching, and other things impose strains and duties which make 'business as usual' impossible, even if we were minded to have it so. Without intensive preparation on the civil front we cannot hope to crush an invader; and without the closest collaboration, before and during the emergency, between civil and military defence we cannot hope for success.

(*d*) Modern Warfare is Lightning War (Blitzkrieg)

This feature in a manner sums up the others. Like world-wide war, it is here, there and everywhere, descending like a bolt from the blue. It goes to war before it declares war. Japan gave the most outstanding example of this: hers was the Blitzkrieg in perfection. As a type of mechanical war, it is a realization of the possibilities of the internal combustion engine installed in an armoured tank or aeroplane, and thereby creating war in new dimensions. As a part of totalitarian war, it strikes at moral as well as physical resistance; the fifth columnist is one of its weapons. It employs infiltration both in the political and military spheres.

Its battle tactic is the delivery of a sharp irresistible thrust, followed by intensive exploitation of the gap created. It searches for soft spots, and these found it rolls up resistance on either side by a fan-like spreading out. The armoured machine leads, unarmed humanity follows.

The purist may object to the use of a German term, but for this there is good reason. The surprise of the late war was the early stage at which a warfare of manœuvre was frozen into trench warfare along a continuous system stretching from Switzerland to the Belgian coast. The weapon which more than any other was responsible for this was the machine gun; and the Germans having realized its characteristics employed it to the full. They became machine-gun minded; and being so minded, they realized the importance of a land arm like the tank, which removed the machine gun's invincibility. Hence, on the basis of the tank they erected a method of warfare which from the lightning nature of its thrust they called Blitzkrieg. It was, moreover, congenial to German ideology; for its sudden thrust of sharp brutality upon the weak spot had its parallel in Nazi methods of rule, and its emphasis on attack restored the old belief in its superiority, which the war of 1914–18 had shaken. In fine, it regenerated in a new medium their classical military teaching.

The handling of a novel master weapon requires preliminary rehearsals under conditions which are a real test. These the Germans obtained by intervention in Spain on the side of Franco, and it was real, even though small-scale, warfare. The Russians, although they fought for the side which lost (though not for lack of courage), learnt in the same way; and when Germany attacked them later, they already had practical knowledge of the Blitzkrieg and what it involved. It was not only the many millions of Russia and the vast areas of Russia which enabled them to fight back so magnificently; it was also their new solidarity of social life, in association with their awareness, through previous training and practical contact, to the realities of the situation. She had early, in fact soon after the late war, developed warfare on the political front, as a preparation for military action; while her army had developed tank warfare and the use of air-borne troops. Thus she was prepared both mentally and materially to meet the new German war machine.

3. The Interdependence of the Military and Industrial Efforts

Later we shall follow the Blitzkrieg on to the battlefield, where the emphasis will be on that side of the time-factor which relates to suddenness. But complementary to this is another aspect, namely the length of time which it takes for the mass production of the instruments which the Blitzkrieg employs to come to full effect. For modern mass production, whether for peace or war, is indirect and roundabout. We do not notice this in peacetime, because then the flow is quiet and continues what has gone before. But war with its new demands brings it into violent relief. First we must build new factories and make the appropriate machine tools. Then we must make the ordinary machines which make the final product. Parallel with this chain of operations is the long interval between original design and full production. Henry Ford's name is associated with this feature of modern industry in its most standardized aspect: we speak sometimes of the Fordization of industry. By 1918 industry had gone a long way in this direction, for we were then in the fourth year of war; and war forces the pace of industrialization.[1] Why were we able in 1918 to replace in a few days the heavy losses of material which we incurred in the final German thrust of March 1918? Because by then and not before we had come into full production. Powers which prepare for war beforehand, which live in a state of 'near war' before they launch into full war, secure an advantage here. For they declare war only when they have got, if not into maximum, at any rate

[1] This was true in different ways of the Crimean War, the American Civil War and the Franco-Prussian War of 1870. The demand for heavier cannon in the Crimean War attracted the inventor Henry Bessemer into the iron industry, and the speedy outcome was Bessemer steel in 1856, the birth year of the age of steel. The demand of the Federal armies for food and transport stimulated railway building and the introduction of labour-saving machinery, especially in agriculture. Canned foods and condensed milk appeared as new industries. In fact, that war may be said to have precipitated America into modern industrialism. The mitrailleuse of 1870 was the first chapter in the history of automatic fire-power: the Vickers machine gun, as finally developed, constituting, both in the fire-power which it delivered as well as in the mode of its construction, a perfect example of the unvarying, standardized, mass repetition so characteristic of modern industry. Similarly in the years before 1914 naval armament both in Britain and other countries stimulated, and was stimulated by, technical progress in metallurgy and mechanical engineering.

into fairly full production. Thereby they reap the harvest of 'total' liaison, i.e. liaison between the industrial and military world. The producers of weapons are the inventors and manufacturers: the consumers the soldiers in battle. The consumer must know the kind of thing which the inventor can design and the manufacturer can multiply, just as the producer must know what the soldier is capable of using. And just as it takes time for an invention to be translated into large-scale fact, so it takes time for a new weapon to develop the tactics of its use. The narrowing of the gap between technical possibility and correct use of the new technique is a prime feature of total liaison. In this connection I have always urged that, if one is asked to look at something new in order to decide whether it is likely to be of value in war, one must never judge its value by what it actually is, but ask oneself rather if it embodies an idea that can be made of value when the necessary adjustments have been made.

Note further the following points:

(*a*) Typically, modern industry is not the building of a few great structures, but the central generation of large-scale power, and by the aid of this power the production, in factories both large and small, of thousands and tens of thousands of relatively few types. The tank and the aeroplane, by comparison with the battleship, are small engines of war; it is their multiplication which gives them such commanding strength.

(*b*) With standardized mass production goes a whole technique of replacement and repair. A familiar example of this is the network of depots which the agricultural implement makers maintain across the prairies of Canada and the American Middle West, each containing the necessary spare parts, which can be rushed up to the farmer in the field without delay during the precious days of harvesting. The mechanized army of to-day must have a similar organization behind it. The quick salvage and repair of tanks in the field may make all the difference between victory and defeat. Libya is offering outstanding lessons, and the work of the Afrika Corps in salvage and repair appears to be of the highest order. Inseparable from this is the question of spare parts. The mere production of large numbers of tanks is of little value unless an ample supply of spare parts is provided also. It is better to have fewer tanks with ample spares to keep them in action than more tanks which, for want of spares, soon become immobile.

(*c*) The problem, of course, has its strategical side, and our Prime Minister put his finger on this when he said recently:

> What are the stages before production comes into flood? The first year, nothing; the second, little; the third, a lot; the fourth, all. Germany was in the final stage at the outbreak of the war; the United States is not yet half-way, though in 1942 their increase will begin to come along with immense momentum. Britain is nearly three-quarters of the way towards the peak of her war industry.

Though our start was dangerously late, it is now reasonably certain that our technicians and factories can deliver the tools. We also know that the Navy (the Royal Navy and the Merchant Navy) will continue their magnificent effort to deliver them to the various theatres of action. We, on our side, must make sure that as fighters on the military and civil front we make the best use of the tools delivered to us. Modern warfare is a new warfare and imposes changes in tactics and organization which reach down to the smallest unit. We must understand, therefore, what the new warfare involves and study it fully and openly. Secrecy over first principles is merely silly. Secrecy over the application of those principles to a situation which requires instant action is sheer suicide. For again I say, and cannot say too often, that if the whole structure of the civil and military defence is not worked out as one plan and is not fully understood, discussed, studied and rehearsed by all those concerned, it will fail us in the hour of trial. If, on the other hand, we have complete collaboration and understanding alike on the material and the psychological side, the invincibility of this island will be ensured.

Part II. MILITARY EVOLUTION

Principal Jacks, reviewing Mr H. G. Wells' *Outline of History*, says: 'Mr Wells would have mankind know where it is before attempting to go any further, and he truly says that the only way to understand where we are is to understand how we got there. That is, to study history as a unitary whole.'

1. The Study of History

Only by studying history as a unitary whole will the student of war arrive at the 'eternal principles' and be able to distinguish these from the 'factors' which are under incessant evolution. We soldiers have often been led astray by confusing factors and principles. Even now, I hear people say that area defence and defence in depth are principles—they are not. They are tactical measures rendered necessary by certain factors, and tactics change as these change. For example, the defence of areas in depth is vital against the armoured attack of to-day, but the linear defence of Wellington, using the fire-power of all available weapons, defeated the French columns, because their narrow frontage precluded the use of more than a very small proportion of their available fire-power. At the same time, when charged by the French cavalry masses at Waterloo, he formed squares and thus adopted the all-round defence of areas which is so necessary to-day.

It is also frequently said that entrenched lines such as we had on the Western Front in the last war are useless to-day. This is not so. Continuous lines of trenches, heavily wired and fully manned, covering the whole area to be defended, would make infiltration impossible, and would nullify, e.g., the tactics used by the Japanese in Malaya. It may be argued that such trenches do not prevent the infiltration of the rear areas by air-borne troops. True, but eliminate ground infiltration and that from the air can be dealt with much more easily, as MacArthur's defence in the Philippines showed.

The modern German doctrine rests on a mass attack of tanks in great depth and on a narrow front. This method, owing to new factors of speed and armour, lessens the number of tanks exposed to the anti-tank weapons of the defence, and by enabling constant reinforcements to take place from behind makes deep penetration possible. But if one was attacking an enemy in the open who had no anti-tank equipment and no depth of defence, it would be better to attack in line and thus to make full use of the fire-power of every tank. Thus it can be seen that tactical methods are merely expedients made appropriate by varying factors. A study of military history from Cyrus the Great to the present day persuades me that there are only four real fundamentals of war and that if we had always kept these in mind we should not have made the many mistakes that we have.

2. The Fundamentals of War

(*a*) The four fundamentals are:

(i) *Fire-Power*. The means by which you destroy your enemy —arrows, bullets, shells, bombs, gas, etc.

(ii) *Mobility*. The means by which you manœuvre your weapons of destruction so that they may be in the right place at the right time.

(iii) *Protection*. The means by which you protect your weapons, and their users, from destruction by the enemy.

(iv) *Morale*. The fighting spirit, which is the means by which the individual is enabled to give his best at all times, and under all circumstances.

(*b*) Fire-power depends on having the best of weapons and being a master in their use. It is the most important of the fundamentals because it is the only one that can, on occasions, win without the help of the other three.

Mobility depends on having the best means of movement for the purpose in hand and using it to the best advantage. It must be conjoined with fire-power and protection.

Protection takes many forms. The best is fire-power, but it may also be given by mobility, armour, static defences, concealment and surprise. Normally it is a combination of all.

Morale comes from being of a fighting race, belief in one's cause, confidence in leaders, confidence in self. The last is essential and is induced by the individual knowing that he is armed with good weapons and understands their use. Physical fitness, too, is important, since tired men lose morale, and therefore care for the physical condition of the troops is a prime factor in the preservation of morale.

But given that equipment is equal on both sides and that one side has greater numbers than the other, then that side will also have greater potential fire-power. The principle is that in computing strengths we must think in terms of weapon-power and correct application of fire-power, and not in terms of numbers of individuals. For, as Ludendorff wrote: 'It is fire that is decisive, not numbers.' The keynote of the last war was the domination of the battlefield by the machine gun, i.e. by fire-power which grew and grew with the ever-increasing number of automatic small arms. To overcome this we produced an enormous quantity of artillery and artillery ammunition, but even this, as noted above, sometimes defeated its object by rendering the ground even more difficult for the infantry to cross. Every change of tactic and material during the last war was designed to overcome the combination of the machine gun, trench and barbed wire. The necessity for cutting the last before the infantry could advance rendered surprise impossible, since it had to be done by artillery action of some duration. This combination remained the master of the battlefield until the tank was evolved and brought to sufficient mechanical perfection to be used in battle. Though the tank was the obvious solution, it had to overcome not only the defences of the enemy but also the hostility of conservative opinion. The war ended too soon for it to succeed in fully overcoming the latter, and that fact is one of the reasons for our failure to develop it as the principal weapon of war in subsequent years. If we soldiers had been imbued with the four fundamentals, we would have realized at once that here was the weapon which for the first time in history enabled the first three fundamentals to be combined in one machine, i.e. fire-power, mobility and protection. True historical study would have revealed that from time immemorial the great captains have been trying to produce this combination—armoured elephants, armoured carts, armoured towers, armoured men and horses, and armoured trains had all been tried, but until the internal combustion engine was invented the problem was impossible of solution. Yet even when the internal combustion engine and the track had enabled the problem to be solved, very few people grasped the fact that at last the master weapon of land warfare had been produced.

3. The Correct Use of Weapons

Whenever a new weapon, or new tactical method, is introduced, it is always looked upon with the gravest suspicion. That is one reason why a study of the History of War is so important, since through it can be seen over the ages the effect on war of such novelties. Such study shows that the human mind has been slow to grasp the possibilities of the new arrival and to adopt the tactics which will put it to the best use. In fact, as in the case of the tank, the new arrival has usually been treated as an adjunct to, and clothed in the tactics of, the older arms.

The story of David and Goliath illustrates perfectly the hard road of the military reformer. Read I Samuel xvii, and you will see his almost invariable lot. First, he is laughed at: if he persists they get angry with him: only if the crisis to be faced is grave enough to cause trepidation amongst the orthodox is he allowed the opportunity of putting his ideas to the test. Even then, like Saul, they will try to dress those ideas in the garments of the old tactics, instead of realizing that a new weapon usually requires a new tactical conception to enable it to develop its full possibility. Even if the new idea succeeds, the successful reformer (who has of course been looked upon as a fanatic) is never able to free himself entirely from that devastating stigma. When it appears that the new idea is finding favour with the authorities and that adherence to it is likely to acquire merit, converts quickly appear; and they, like converts to a new religion, forgetting their former derision and opposition, soon become more fanatical than those who evolved the idea.

Do not imagine that this anti-change mentality is a peculiarity of age. Not at all. My experience is that it is often harder to convince arrogant youth than cautious old age. Youth is so desperately certain, whereas old age has seen so many things go wrong, seen perhaps two generations of youth completely wrong, that it is not so desperately certain and therefore easier to convert.

Those who advocated the use of tanks experienced at first the difficulties of David. The orthodox conceived them as a mere adjunct to the infantry, and as such they were used until the tank people got their own way at the Battle of Cambrai on 20 November 1917. The result was an unprecedented success. The penetration was far deeper and the casualties proportionately fewer than in any previous battle. So little had this success been expected by the Higher Command that hardly any preparation had been made to exploit it. For this reason the final results fell far short of those warranted by the initial success. Be that as it may, Cambrai marked an epoch in the history of war. Armoured war had arrived, the Blitzkrieg was born, and in one respect at any rate it was orthodox. Twenty-one years later it reached manhood.

From Cambrai onwards experience in the use of tanks always taught the same lesson, namely, that when used in 'penny packets' as an auxiliary to the infantry they were seldom successful, but that when used in large numbers on suitable ground and therefore with all available support organized primarily for them, their success was phenomenal, as at Cambrai, Hamel, Amiens, etc. It was now realized that as many tanks as resources permit must be used in a break-through operation and that the mass attack of tanks must be aimed at. Even an infantry and 'I' tank attack must be staged with tanks as the main assault arm and the infantry as the 'moppers-up' and occupiers of the ground captured by the tanks. Here then is seen the fundamental principle on which the modern German attack is based, written by the end of the last war, made plain 'that he who runs may read'. So much for the lessons of the last war being inapplicable to this one!

Other features of the modern German methods can also be seen emerging from the experiences of 1917–18. Thus at the Battle of Amiens, 8 August 1918, machine gunners were carried forward in tanks so that they might hold forward positions against counter-attacks, until the arrival of the main infantry forces. Again, at this time squadrons of aircraft were told off for special co-operation with the tanks, and thus we see the emergence of that close co-operation between the fighting machines in the air and those on the ground which is such a striking feature of the Blitzkreig. Finally, in the summer of 1918 the tentative planning began for the great battle to be fought in April 1919, which it was hoped would win the war. The area of attack was to be selected with reference to the suitability of its terrain for tanks both as regards the initial attack and its subsequent exploitation. The battle was to take the form of breaking through the German defences with a normal set-piece attack by artillery, heavy tanks, infantry, etc., after which was to pass through the gap thus made a large force of light tanks, supported by aircraft and supplied by cross-country tractors. These were to attack headquarters, communications, supply and, in fact, the whole rear installations of the German army and thus throw all into confusion. The tank and air offensive would, of course, have been exploited as rapidly as possible by the rest of the allied armies. For this battle some thousands of tanks and cross-country tractors were ordered in 1918 as well as the countless other munitions of war required for an offensive on such a scale. It is thus clear that, except for air-borne troops, the whole idea, untried and undeveloped to be sure, of the Blitzkrieg was in being in 1918. The most modern method of modern war is the product of British brains.

Before leaving the last war there is one observation I would make, and that is that the factors with which the Germans had to deal in their attack on France in 1940 were very different from those which would have confronted them if they had tried a similar attack in 1917–18. If in 1914, i.e. at the beginning of that war, they could have produced an armoured attack through Belgium, the result of 1940 would have been anticipated. Unprepared and unequipped to meet such a form of attack, the French and British armies of 1914 would have succumbed as did their successors of 1940. But by 1917 the Germans would have found a complete system of deeply trenched fortifications, defended in considerable depth by well-armed troops and stretching from Switzerland to the sea; and it would have been just as necessary for them to break a gap in that system by a normal 'set-piece' attack, as it would have been for us in 1919, because the majority of tanks in use to-day could not have passed through that system until trenches had been filled in or bridged and the way made clear for the exploiting tanks to pass through to the back areas. But once the break-through had been made, and with sufficient suitable assault tanks it could have been made, it is difficult to see what could have prevented their complete success. For there would have been no armoured forces to oppose them, and the only true answer to the armoured attack is the armoured counter-attack.

It suffices to say in conclusion that to employ new weapons correctly and to appreciate properly the new factors necessitates a change of ideas. Ideas are the hardest things to change. The change of ideas, that is of mental outlook, has been invariably the hardest task of the reformer, and that is why I contend that the mental is more important than the material. If you have the right idea you will get there in the end, however many mistakes you make on the way. But if you start with the wrong idea, you will never get there. That right idea can only be obtained from experience, and the mass of experience that lies recorded in history can only be obtained by a correct study of history. The future is always a result of evolution from the past, and therefore to have any hope of forecasting future developments one must know the principles underlying the evolution of the past as well as the developments of the present.

4. What the Fundamental of Mobility implies

Just as the last war gave us the beginnings of mechanized warfare, so the present one developed these into a code of war. Therefore, before dealing in detail with the German doctrine of the mechanized attack, I will indicate a few of the characteristics of the new code.

The troops employed in this form of war may be divided into (i) mechanized troops, which are those organized for battle on a mechanized basis (they may be armoured or unarmoured) and (ii) motorized troops, which are those temporarily motorized for a particular purpose. I have said that the German practice in the mechanized attack is to have as its spear head a mass of tanks, organized in great depth and on a narrow front. But they do not advance to the battle in mass formation. On the contrary, they have adopted for their approach the Napoleonic method. Before the Napoleonic era strategical advance and tactical manœuvre were both of them concentrated. Napoleon taught dispersion in approach. He moved his armies on a wide front, concentrating them only for the battle. Such dispersion increased the mobility of the force, since it eased congestion and facilitated movement and supply. The whole advance was covered by a screen of mobile troops, who kept touch with, and observation on, the enemy, while hiding from him the movements of the approaching columns. This method also tended to confuse enemy commanders, since they could not be certain against what part of their army Napoleon would concentrate. The Germans are applying the same principle in even greater degree to mechanized forces, for the more mobile the force, the wider the front on which it can operate through its power of quicker concentration. Modern movement must be considered in terms of time and not of distance. In addition, the greater the dispersion and the less vehicles there are on the road, the less is the congestion and the less is the target for air attack. The dispersion must be in depth as well as lateral. Columns of mechanically propelled vehicles must be kept farther back than the old slow-moving columns, so as to allow room for manœuvre. They must not be committed too far forward, until the best route is certain.

Greater mobility than your enemy gives you the choice of ground, the power of surprise and the ability to anticipate. The Mongols, greatest and most mobile conquerors of the past, selected the ground for their battle, and manœuvred in order to secure it. With mechanized forces the same idea applies. Manœuvre should be used in order to secure the most suitable ground on which to fight.

Increased mobility adds to the requirements of mental training, for every increase in potential speed means that we must train our minds to move at the increased speed involved. We must think more rapidly and also farther ahead. As we have to think more rapidly, so we have to ensure that our decisions are conveyed more rapidly to those who must put them into effect. In the fast-moving mechanized battle, there is no time for detailed written orders. Verbal orders and wireless are the key to the rapid action of mobile forces.

During the advance to the battle, and during the battle itself, the mechanized forces must be supported at all times by the air arm. As I have written on many occasions during the period between the last war and this, 'it must always be remembered that the machines in the air and the machines on the ground are two parts of one whole. Neither are complete without the other. For tactical and strategical operations they should definitely train together.' Therefore there must be complete liaison and inter-communication throughout the whole action between the air and the ground. This has been brought to a high pitch of perfection in the German Army.

I want it to be quite clear that I am no advocate of the Army having its own separate air force for this purpose. I am a great believer in the principle of specialization, and I consider that the R.A.F. with the splendid organization it has developed can produce the requirements of the Army, and train people in the correct use of these, better than the Army can. To produce the concerted action required the first essential was for the Army to know exactly what it wanted and why, and then to insist on the R.A.F. providing it. But I fear that on this issue the trumpet of the Army emitted a most uncertain sound, and I am not surprised that at the outbreak of the war a proper working arrangement had not been reached. Moreover, we have not yet got sufficient aeroplanes for all purposes, nor sufficient different types for all requirements, and until we have, their allotment can only be decided with reference to the requirements of the war as a whole. If we tie up our various requirements in water-tight compartments, we shall make more difficult the proper distribution of our resources as the occasion demands.

While the armoured mass forms the spear-head of the attack, it must be followed by unarmoured troops, to open out the breach made by the tanks, and thus to make a road through the enemy defences which can be used by the rapid-moving pursuit troops, who will pour through on the heels of the armoured forces and take up the action in the area beyond the first systems of enemy defences. Therefore again the liaison and communication between the armoured and the unarmoured forces must be of the closest nature.

The foregoing shows the necessity of team work: the armoured troops and the unarmoured troops must, in this attack, work as one team, on a definite plan most carefully worked out beforehand. At the same time, that plan must only constitute a general direction, to which all conform, and must in no way hamper individual initiative; for the essence of mobile war is the immediate exploitation of any weakness shown in the enemy's system. That is why in the German Army of to-day such tremendous attention is paid to decentralization, and to imbuing the lower commanders, officers, N.C.O.s, and even privates with the idea of initiative and immediate action. In fact, as might be expected under such circumstances, command and control are decentralized on the principle that the man on the spot can far more easily judge what action should be taken, and must be prepared to exercise his judgment and use his initiative. Very great stress is laid on the need for initiative by even the most junior commanders. Truppenfuehrung—the equivalent of our Field Service Regulations—states 'mistakes and even disobedience are better than inertia', and again 'better a faulty plan which shows boldness and decision than a perfect plan enmeshed in uncertainty'. In fact, there would appear to be only one thing that constitutes an unforgivable sin in the German doctrine and that is to do nothing.

I am a firm believer in the principle of decentralization both in the civil and military spheres of our war effort. For any operation a central plan must be made, but such plan must not lead to centralization in the hands of officials. No. The secret of success lies in the modern German principle of decentralization. Central authority must make a plan, but the implementation of that plan at the lower levels must be left to the initiative, common-sense, and local knowledge of 'the people on the spot'.

Part III. THE PRESENT AND WHY

And the Lord was with Judah; and he drave out the inhabitants of the mountain; but could not drive out the inhabitants of the valley, because they had chariots of iron. Judges i. 19.

Unlike Nelson's day, now no valour can compensate for mechanical inferiority. Admiral Lord Fisher.

1. Lord Gort's Despatches and their Lessons

(*a*) These despatches brought to light a large number of deficiencies in equipment and training. He points out that while the enemy concentrated not less than ten armoured divisions for his attack, the B.E.F. had only seven cavalry regiments equipped with light tanks, one regiment of armoured cars of an obsolete pattern, and two battalions of infantry tanks, of whom the majority were armed only with one machine gun. Thus the B.E.F. was, as he says, inadequately equipped for the task it had to fulfil. In addition, he states:

'The days are past when armies can be hurriedly raised, equipped and placed in the field, for modern war demands the ever-increasing use of complicated material. Indeed the scientific side of warfare has been evolving at a very rapid rate ever since the end of the war of 1914–18 and is continuing to do so. Modern equipment requires time to design and produce, and once it is produced, further time is required to train troops in its technical and tactical uses. Improvised arrangements, made at short notice, can only lead to the shortage of essential equipment, the production of inferior articles, and the unskilful handling of weapons and vehicles on the battle-field.'

(*b*) Many trenchant criticisms appeared in the press after the publication of these despatches. I will only quote three of them:

(i) 'The thing which will hit any careful reader most forcibly is the fact that when the British Expeditionary Force went to the continent those who sent it, and that applies to the military as well as the political leaders, either had no conception of modern war or were apparently perfectly powerless to translate their conception of need into reality.'

(ii) 'From a military point of view there was nothing wrong with the decision to advance into Belgium; what was not just wrong, but criminal, was to advance with 1914 troops to meet the German 1940 military machine.'

(iii) Lord Gort in an appendix on the lessons of the campaign points out the paramount importance of equipment—'but the paramount importance of equipment was visible in every page of German technical military literature and was part of the common knowledge of any serious military student'.

(*c*) Why then did we send such an inadequately equipped B.E.F. to meet a German army whose equipment and tactical ideas were well known to British students of armoured war? It appears all the more extraordinary in the light of what I have already told you of the state of affairs, mental and material, in 1918. We had at that time, as I have already shown, the necessary experience to lay our preparations for modern war on a sure foundation and thus, during the years that Germany was not allowed to re-arm, to forestall her in her preparations.

(*d*) Recrimination and criticism there have been in plenty and I have no intention to add to them, so I will merely emphasize a few fundamental facts as a warning to those in whose hands may lie responsibility for our future military preparedness between this war and the possible next, for we must face the fact that there may be another war, however hard we may strive to prevent it.

2. The Reasons Why

For if the trumpet give an uncertain sound, who shall prepare himself to the battle? I Corinthians xiv. 8.

(*a*) The first culprit is the British public, you and I, the ordinary citizens of this country. Now it is the fashion to blame all the political parties that held power during the last twenty years. But do not forget that we ordinary citizens elected those governments. In one periodical I read it was put as follows:

> 'The public must not forget that the support which successive governments enjoyed was after all of their giving, and that the critics—of whom Mr Churchill was one of the most notable—had very little of that support, and were regarded as political playboys at the best or dangerous mischief makers at the worst. Any party which had braved popular indignation in those days by facing realities (i.e. by advocating spending money on preparation for war) would have earned great merit and future

power, but would never have been in office during the years between the wars. There is very little to choose between the conduct of the public and the politicians. The fact is we went through a very bad phase and had the very governments the nation wanted.'

As a well-known M.P. said to me a short time ago:

> 'I, with many others, am greatly to blame. We could not believe that after the last war, with all its horrors, anyone would make war again. We were obsessed with Geneva and believed that the only hope for peace was in disarmament, and that therefore any arming which we did in the way of modern equipment would militate against the fulfilment of the ideal which we had at heart. We now see clearly how wrong we were, for how could we expect support for the League, and all it stood for, from the smaller nations unless we were strong enough to protect them from the aggression of a recalcitrant power? We realize, now that it is too late, that the only way to keep the peace is for the peace-loving nations to be strong enough to enforce it.'

I think that statement would sum up the feelings of the average citizen in so far as he thought about the matter at all. But we must remember that there are a very great number of us who do not bother to think, and who, in the truly British manner, refuse to believe anything that they do not want to believe.

The fact remains, however, that the Army was cut down, money for new equipment was not forthcoming, and that while a faithful band of armoured-war enthusiasts tried to keep the flag flying, their doings were not considered of paramount importance and little money was given to them for development and for that essential experiment without which development cannot take place.

Much could be said on this matter, but I will only quote from Mr Bevin as reported in *The Times* of 5 December 1941, when referring to the use of skilled men by the Army:

> 'But, on the other hand, he agreed that the Army must not waste these men. His sympathy was very largely with the Army in this respect. Half of the criticism of our Army in wartime was due to the fact that we had agreed to save 6*d*. on the income tax by breaking it up in peace time. That was one of our big

troubles, and when the war was over he hoped that we should remember it. He would never be a party to such a policy again after what he had seen in this war. He knew industrialists pretty well and he knew their expert ability. If the same enquiry was made about misfits in industry as was made in regard to the Army he thought that just as many misfits would be found in industry as in the Army.'

(*b*) Now what of us soldiers? It is true that we did labour under great difficulties, but can we say that we are blameless in this matter?

I do not think that we can, and I believe that the main reason was that a very large number of soldiers were not absolutely convinced that the tank was the master weapon of ground warfare, and were not prepared to admit that in any major war all else would depend on air and tank superiority. This does not mean that the tank, or the aeroplane, can win wars without the assistance of the other arms. But it does mean that without adequate armoured forces and air support, the other arms are quite ineffective and that therefore armoured war is the essence of modern war, in which the unarmoured troops must co-ordinate their efforts to fit in with the armoured, and not vice versa. I feel that if the military leaders had really felt as fanatically certain on this subject as did that small band of armoured-war enthusiasts of which I had the honour to be one, then, while no large armoured force could have been formed, a greater measure of progress would have been made and a far better foundation laid on which to build when the money became available. I was personally very much in this business and I could write a great deal about it, as to what could have been done and was not done, but suffice it to say that if we soldiers had grasped the obvious teaching of the past, we should have felt sure of the right developments of the future, and with that certainty in our mind our trumpet would not have given an uncertain sound and we should have striven more whole-heartedly and effectively for those things in which the war found us so lamentably deficient.

(*c*) A study of the last war had left those of us who believed that in the tank we had at last discovered the major weapon of land warfare with an intense desire to experiment in, and develop, what we called 'armoured war'. It was obvious that whatever form the next war took the land part of it would be dominated by the motor vehicle in every shape and form.

It was also certain that in no war would we be likely to be able to take the offensive at the beginning, and that therefore our actions would perforce be defensive until our offensive power had been developed. Therefore the main object in view was to produce mechanized forces strong enough to permit us to support such a generally defensive strategy with adequate counter-offensive action. At the same time our experiments and developments must lay the foundation of that powerfully offensive army without which victory can never be attained.

To this end I put forward in 1924 a proposal for the formation of an experimental mechanized Brigade of Armoured and Unarmoured Units. I fully recognized at the time that this Brigade, in the form suggested and with the equipment available, was not suitable for modern war. I felt convinced, however, that if only a completely mechanized force could be brought into being, and experimented with, the mental effect of its existence would soon lead to an evolution of ideas, materials and organization, that would in due course lead us to the desired goal.

A force of this nature was formed on Salisbury Plain in 1927 and, in spite of the difficulties, political, financial and military, which I have already touched upon, it did lead to a considerable advance and eventually to the Armoured Division as it exists to-day.

There is no time to go into any details of this development, so it must suffice to say that while advance was made, that advance fell far short of what might have been achieved: the main reason being that those concerned never had the financial support necessary to carry out the experiment on a sufficiently large scale either as regards work in the field or as regards the design and production of new armoured fighting vehicles. Nor was combined action with the R.A.F. sufficiently explored.

The Germans, on the other hand, adopted whole-heartedly the doctrine of armoured war. Their trumpet gave no uncertain sound and those who listened to it had no doubt of its meaning. This, and their experience in Spain, Poland, France and the Balkans, gave them in advance tactical and technical training which our own armoured forces did not possess when they had to meet the picked personnel and material of the Afrika Corps in North Africa.

We fanatical adherents had no doubt of what the Germans were doing, of their aims, organization and tactical methods. Just as Hitler has never concealed his objects or his methods, so the German Army never concealed to initiated observers the principles to which they adhered or the tactical methods which they proposed to adopt. The wars on the political and military fronts were all part and parcel of their doctrine, the former merely preceding and later coinciding with the latter. But we had not behind us the support to produce the results which they produced, and so the Germans perfected the technique of armoured war which we initiated and could have perfected if we had been given their opportunities.

Chapter II

THE EMPLOYMENT OF ARMOURED AND MECHANIZED FORCES IN THE OFFENSIVE

Owing to the effect of tanks, our operations on the Western Front have now practically assumed the character of a game of chance. The General Staff can no longer work with definite factors.

Statement by General Ludendorff on 30 September 1918 at a State Conference.

August 8th was the black day of the German Army in the history of this war.

The Divisions in line...allowed themselves to be completely overwhelmed. Divisional staffs were surprised in their headquarters by enemy tanks....Mass attacks by tanks and artificial fog remained hereafter our most dangerous enemies. Ludendorff, *My War Memories, 1914–18*.

1. General

(*a*) Any formation which is designed to operate independently, or even in a semi-independent role, in a major operation must combine in itself the characteristics of all arms. Therefore an Armoured Division must be organized on this principle, and this is the case both with the British Armoured and the German Panzer[1] Divisions.

The exact composition of such divisions alters from time to time as rendered necessary by experience of tactical requirements, and by the factors which have to be dealt with in different campaigns or periods of a campaign.

Therefore all that can be said is that such formations consist of three groups:

(i) *The Reconnaissance Group*, which normally is made up of Motor-Cyclists, Armoured Cars, and Unarmoured Mechanized Troops.

(ii) *The Armoured Group*, consisting of Tank Battalions. These in the British Division are organized in Brigades, and in the German in Regiments, which are approximately the same thing.

[1] Panzer is the German word for 'armour'. Thus, Panzerkreuzer = 'armoured cruiser', and so on.

At the beginning of the war a Panzer Division had about 450 Tanks in its Armoured Group.

(iii) *The Support Group*, consisting of Mechanized Infantry, Medium and Heavy Machine Guns, Anti-Tank Guns, Artillery (including Anti-Air), and Royal Engineers (or their equivalent), and the various services.

(*b*) There are two main roles in which armoured and mechanized forces can be employed:

(i) Tactical co-operation with the main army on the same battlefield.

(ii) Strategical co-operation in the same theatre of war, but not necessarily on the same battlefield.

(*c*) On a major battle-front, when large armies are engaged, strategic co-operation may not be able to play an important part, the flanks are not open, the defence is deep, the battle is closely joined, and the whole area of fighting is, in a sense, one great battlefield.

So all I will say about strategic co-operation is that, providing that the theatre of war suits it, an armoured and mechanized force might be employed to operate, in conjunction with air forces, against the line of communication and supply of an enemy which is already being engaged by the main army on a battle-front a considerable distance away from the area in which such mobile forces are operating.

In the last war the Mesopotamian and Palestine Campaigns are examples of the possibilities of such an operation.

Given open flanks, suitable terrain and adequate mobile forces, such an operation will always present the opportunity for decisive action against the enemy. Such an opportunity should not be lost sight of in the type of war that may develop in the vast area from Persia to North Africa.

The question of supply, especially by air, will be a problem that will require very careful arrangements. Air supply for Mobile Forces is a matter which I experimented with on a small scale in Iraq in 1922, and which I believe to be the answer. The Germans have already shown us its great possibilities on more than one occasion.

Remember that in every phase of tactical and strategical operations with armoured and mechanized forces, the machines in the air and the machines on the ground are two parts of one whole. Neither is complete without the other.

(*d*) In tactical co-operation armoured and mechanized troops will normally have five main roles to fulfil:

(i) Close tactical co-operation with the unmechanized troops in the initial break-through, if the enemy position does not permit of initial rupture by an armoured force.

(ii) Semi-independent tactical co-operation in action against the enemy gun positions, immediate headquarters, reserves and communications.

(iii) Independent tactical co-operation in action against more distant objectives, such as rear headquarters, lines of communication, aerodromes, supply centres, and in cutting all communication between the enemy troops in the battle and their home bases.

(iv) Defeat of the enemy's armoured counter-attack.

(v) Close tactical co-operation in the 'mopping-up' of the enemy's defensive position after the main break-through has been effected.

2. Co-operation between Armoured and Unarmoured Troops

(*a*) Close co-operation between armoured and unarmoured troops is always difficult, as the methods by which they operate are essentially different.

For example, small arm fire, especially that of automatic weapons, is the chief danger to unarmoured troops. The chief characteristic of the tank, on the other hand, is that it is impervious to such fire, and its danger comes from artillery and special anti-tank weapons. This asset is most marked when tanks have only to pass through a zone of hostile fire incidentally and so can use their speed, and latitude of manœuvre, on the way to an objective beyond. When the task of the tanks is to assist unarmoured troops through such a zone, they must, perforce, move more slowly and seek out and silence that chief enemy of the unarmoured troops, the hostile machine gun.

In this process they will inevitably expose themselves to their own chief enemies, the artillery and anti-tank weapons, far more than if they were operating independently of the unarmoured troops.

Thus the close co-operation between armoured and unarmoured troops has inherent difficulties, which are greatly enhanced when the action is not being conducted as a co-ordinated operation in which artillery, machine guns, infantry and infantry tanks work together on a definite plan.

(*b*) In the British Army we have, in addition to armoured divisions, brigades of 'I' (or Infantry) tanks which are specially designed for the close tactical support of infantry when the latter are engaged in close fighting, such as attacks on fortified positions, mopping up in the wake of armoured divisions, etc. Normally these 'I' tanks have heavier armour, less speed and greater obstacle-crossing capacity, than those designed for work with fast-moving armoured and mechanized formations.

Therefore, tanks which are not designed for the close tactical support of infantry should not be used for this purpose. Tanks of the fast cruiser type are unsuited for it by their design and their use for it will result in heavy tank casualties.

Tanks, like warships, are designed for different tactical purposes, and it is most uneconomical to use them for purposes for which they are not intended.

(*c*) When armoured forces are employed on distant strategical or tactical roles, the unarmoured mechanized troops will assist them by reconnaissance, by seizing and holding crossings over obstacles so as to enable the armoured troops to cross such obstacles without fighting, by clearing up minor enemy resistance so as to ensure their line of retreat when their mission is accomplished, by protecting advanced landing grounds and supply arrangements, and by forming mobile tank harbours from which the armoured troops can operate. This last is a most important consideration, and armoured forces must always be followed in the advance and preceded in the retreat by such mobile harbours, wherein they can find protection while replenishment and repair proceeds, and where some rest for the crews can be obtained.

(*d*) The armoured attack will normally, in varying degrees, be confronted with:

(i) Fluid conditions in which the enemy has not had time to

'settle down' and prepare a defensive system in strength and depth.

(ii) Similar conditions caused by the magnitude of the area of operations, as in North Africa, and by the forces being so small in relation to the area, that it is impossible to prepare a defensive system of such depth and extent as will enforce a frontal attack against some portion of it.

(iii) Conditions in which the enemy disposes of sufficient man and weapon power to organize his defence in such extent, and in such depth, as will enforce a frontal attack on some portion of it and thus a considerable period of close fighting before the system can be broken through and conditions approximating to open warfare can be attained.

In cases (i) and (ii) an armoured attack carried through in great strength and at the highest possible speed will normally be the best solution. The enemy will not know exactly when or where the attack is coming, and the armoured forces will probably be through him and into his rear areas in great and concentrated strength before he can carry out an armoured counter-attack. If the attack can be carried out at two or more points in the enemy's front at approximately the same time, it will produce the best and quickest results. The advisability of this, however, is conditioned by the relative tank strength of the attacker and the attacked.

The main principle to be adhered to is that the defender's armoured troops must be destroyed when they are encountered. As soon as the attacking tanks are through the defences, they will force the defender to concentrate all available tanks against them. The attacker must meet that counter-attack in the greatest possible strength, and he must run no risk of being beaten in detail through having divided his armoured forces in the attack. In circumstances such as we are now visualizing the whole essence of success is the destruction of the defender's armoured forces. Once that has been done the battle will resolve itself into a 'mopping-up' operation. It is in these conditions that the Blitzkrieg is so successful.

Case (iii), however, i.e. the attack on a highly organized system, presents, in my opinion, quite a different picture and one which requires quite a different solution if the attack is to achieve permanent, instead of temporary, success.

The commander of an armoured force whose objective is beyond the enemy's main defensive system will wish to keep his force as intact as possible during its passage through this system, and to expend the minimum of ammunition and petrol on the way through. Therefore, his object will be to get through with the least amount of serious fighting. If he has to fight his way through he will expend ammunition that he requires for his real task, and he will risk a loss of tanks and a disorganization of his force which will prejudice the full performance of that task. Therefore a way through must be made for him in order that his passage may be as free as possible. This entails a 'set-piece' attack of some kind, varying in its degree according to the strength of the defensive system, and the fighting qualities and morale of the enemy.

In considering an operation of this nature it is best to take the case of a break-through attack which requires the maximum of effort to effect, and which approaches most closely to the 'set-piece' attack as we knew it in the last war. If we do this, we are then in a position to visualize any form of attack from those necessitated by the conditions outlined in (*d*) (i) and (ii) above to that which we will now consider.

3. General Picture of a Break-through Attack where the enemy position is of such strength that a sudden rupture by a massed armoured attack cannot be accomplished

(*a*) Whether the operation be made by a force acting on the offensive, or whether it be in the nature of a counter-offensive, if the operation is intended to be a decisive one its general form will be the same.

(*b*) To get a clear picture in our minds of the form such an operation will take, the battle, however long its duration, should be considered as consisting of five phases. Such phases may follow each other rapidly in one almost continuous operation, or take place in successive stages, with an interval of time between them.

It must be borne in mind, however, that the less the interval that occurs between any two phases and the greater the speed with which the whole operation is carried through, the less time has the enemy for organizing counter-measures, the less long are the attacking troops exposed to the enemy's fire, the less casualties will they suffer, and the more chance there is of ultimate success. Therefore, the ideal will be to carry through the phases in one continuous action, and with as little delay between them as possible. The degree, however, to which this ideal can be approached will depend on the resources available, and the individual circumstances of each case.

The phases referred to are as follows:

Phase 1

The capture, and occupation, of the main enemy position, i.e. the area covered by his defended localities.

Phase 2

The overruning of his forward gun positions, and occupation of the area immediately behind his main position, in which are located the forward centres of command, communication, reserves and supply.

Phase 3

(i) The advance of an armoured force against objectives immediately beyond the area occupied in Phases 1 and 2.

(ii) The protection of the flanks of the break-through in that area, and the mopping-up of the objectives overrun by the tanks.

(iii) The widening of the gap by action against the enemy flanks and rear, if required.

Phase 4

(i) The advance of an armoured force against such distant objectives as may be given to it.

(ii) The advance of the mechanized unarmoured troops, to support the armoured force in its action against such distant objectives.

Phase 5

The general advance.

(*c*) If it is intended to use armoured divisions, whether with mobile troops or not, to exploit success, the point of rupture must be chosen with a view to allowing such divisions to debouch into country suitable for their employment. In addition, the ground over which the main attack takes place must be such as to enable it to be put through with the minimum of delay as soon as the tactical situation permits of its advance.

In this connection it may be desirable on some occasions to put the armoured divisions through in Phase 2, instead of in Phase 3.

If, however, the objectives of an armoured division lie some distance behind the enemy's defended area, it is obviously undesirable to make its first task the overrunning of the enemy's artillery positions, as losses in tanks and expenditure of ammunition in this stage of its operations will handicap it in its later tasks.

(*d*) The inherent difficulties of employing armoured and unarmoured troops in close co-operation have been explained. The fact is that the characteristics of both arms are so different that the same piece of ground is unlikely to be equally favourable to both.

(*e*) Given favourable opportunities, the tank is the most powerful of assault arms. Therefore, when infantry and tanks are being used in close co-operation in the attack:

(i) The area of attack must be chosen from the tank point of view.

(ii) The assault will be a tank assault; and this presupposes attack in depth by the tanks.

(iii) All available covering fire, by artillery and machine guns, will be used to assist the tanks.

(iv) The infantry plan must be to occupy, and mop up, the area conquered by the tanks.

(*f*) The tank effort must be a continuous one staged in depth, each echelon of tanks going straight for its objectives, and making use of ground and manœuvre to turn the conditions to their advantage as much as possible. The depth to which the leading echelon of tanks will go depends on the frontage of the attack, the nature of the ground, the number of tanks available and the nature of the enemy defences.

One echelon of tanks will always keep in close touch with the leading infantry, and render them close physical assistance throughout their advance.

(*g*) Thus, the ideal would be that at the crisis of the battle, that is to say when the infantry are entering the main defended zone, the whole tank strength should be in action, attacking the enemy in every portion of the area of attack from his front line to the back of the area containing his supporting artillery and his main forward centres of command and supply.

In this manner, the tanks are enabled to make full use of their mobility, and the maximum assistance is given to the infantry by the depth of their attack, which thus disorganizes the whole of the defensive arrangements at the most critical time from the infantry point of view.

4. Some General Principles regarding the Employment of Armoured and Mechanized Troops in an operation of this nature

(*a*) I do not intend to go into details of the co-operation of tanks and infantry in the attack, or that of the supporting fire of artillery and machine guns.

There are two general points, however, on which I must lay particular emphasis:

(i) The co-ordination of the effort of artillery, machine guns and tanks is of primary importance. The modern battle is won by fire and not by numbers of unarmoured men. Therefore, the combination of the static fire power of artillery, machine guns and mortars, combined with the mobile fire-power of tanks, is the best method of ensuring that the infantry can occupy their objectives with the minimum of loss.

(ii) Surprise is of great importance in any operation of war, but in tank operations it is quite vital. Initial surprise, exploited to the utmost by speed of action, and by dealing the enemy blow after blow with no respite for recovery, is the method which offers the best chance of decisive victory. Cambrai in 1917 and Poland in 1939 offer strong examples of this.

(*b*) In Phase 3 (i) of the attack the object is to paralyse completely the enemy's power of command, reinforcement, communication and supply in the area immediately behind the battlefield. In other words, to paralyse the enemy's control during the crisis of the battle, when the break-through having succeeded and the main gun positions opposite the point of rupture having been overrun, the attackers are widening the gap, expanding their area of attack, and rolling up the enemy defences by attacking them in flank and rear.

(*c*) The armoured divisions having been passed through immediately on the completion of Phase 2 should find themselves fresh and intact in the area of enemy divisional and corps headquarters. On the way to this area they should avoid, by their speed and power of manœuvre, becoming involved in major fighting, and should make direct, and with all speed, to the places, or areas, allotted to them.

(*d*) The allotment of their objectives will depend on the information available before they are launched, and throughout the operation touch with army headquarters should be maintained by wireless so that the latest information about the enemy will be supplied to the commander. His headquarters will itself be in touch with the air forces detailed to assist the operation, and will be receiving information from them.

(*e*) The unarmoured mechanized troops will be passed through in the wake of the armoured divisions, as soon as circumstances permit. Once all the mobile forces, armoured and unarmoured, have reached the area assigned to them, the commander will allot to them their respective tasks. It may well be that he will divide his forces into several different columns, each comprising both armoured and unarmoured troops, and allot them separate tasks. What he does will depend on the circumstances of the case, and the resources at his disposal.

Although nothing definite can be laid down on this point, it is obvious that the greater the number of enemy headquarters, etc. which are attacked at once, the quicker will the paralysis mentioned in paragraph 4 (*b*) (p. 36) be brought about.

The unarmoured troops, having greater man-power than the armoured, will complete the destruction, or occupation, of the various objectives, and will act generally as outlined in paragraph 2 (*c*) (p. 31).

We have had an example of such columns in the 'desert commandos' formed during the operations against General Rommel's army in North Africa. These commandos were highly mobile columns, of varying size, made up of some tanks, some armoured cars, field, anti-tank and anti-aircraft artillery, and motorized infantry. Each vehicle was self-contained with rations, food, water and petrol for several days. They could move over the desert at from 20 to 30 miles per hour. They got round the enemy's rear, harried him, and did him a great deal of damage.

(*f*) In Phase 4 the mobile forces must aim at operating in such an area as to block the exits from the battlefield, thus preventing the escape of the hostile army, while also preventing all reinforcement to it.

(*g*) The tactics of the mobile troops, in brief, will be to use their speed and power of manœuvre, combined with their wireless intercommunication, in such a way as to disperse over a large area, attacking many enemy installations simultaneously, surprising the enemy by their very ubiquitousness, and yet retaining the ability to concentrate rapidly if required to do so by the commander.

(*h*) Further, there is the possibility that motorized troops from the main army may take over the areas occupied by the mobile forces. If this can be effected, the mobile troops thus released will be enabled to play their full part in Phase 5, namely the pursuit and destruction of such of the hostile forces as still exist in fighting formations.

(*i*) The commander must at all times be able to concentrate sufficient tanks to meet any attack from enemy armoured forces that have not yet been destroyed. This seems to be emphasized by recent experience in Libya.

The foregoing is, put very briefly, the general teaching on this subject which I was giving in 1939.

I also taught that the initial break-through should normally, against a well-organized defensive position, take the form of a 'set-piece' attack, in which infantry and infantry tanks, supported by artillery, machine guns, etc., made a gap in the enemy's defensive system: the armoured and mechanized forces then passing through the gap thus made.

In spite of what the Germans have achieved I believe that, where large armies are opposing each other, where the morale of the defenders is high, where the defensive system is of adequate depth, and where tank-obstacles and minefields are part of the defence, a break-through with tanks on the German plan will not produce disintegration, and will not result in a collapse of the defensive organization.

Therefore, I still believe that breaks-through which do not ensure that the path behind the armoured divisions is not only properly cleared, but also effectively occupied, will render the work of the armoured divisions ineffective and, if persisted in, will eventually lead to defeat.

When details are to hand I believe that this view will be confirmed by the German experiences in the Russian campaign.

5. Aspects of the Modern Attack restated and summarized

Pre-war teaching and present war experience enable us to summarize the principles of the modern attack in the following terms:

(*a*) It should be launched before the enemy has had time to 'settle down', and must be a surprise. Surprise is of great importance in any operation of war; in armoured war it is the essence of its technique. Initial surprise, exploited to the utmost by speed of action and by dealing the enemy blow after blow with no respite for recovery, is the only method by which decisive victory can be obtained.

(*b*) It must envisage the concentration of every available resource of men and material at the decisive point, or points, where the rupture is intended.

(*c*) It must be launched, if possible, against a point in the enemy defence which enables the attackers to take his defensive organization in flank.

(*d*) It must aim at breaking through on a comparatively narrow front, but such break-through must allow the enemy's defensive organization on the flanks of the break-through to be attacked in flank and rear.

(*e*) It must be selected from the point of view of the tank attack, the ground being suitable for the tanks which carry out the initial rupture, and also for the tank formations which exploit the initial success.

(*f*) It must lend itself to the fire-plan, if possible enabling the attack to be supported by enfilade fire after the first break-through has been effected.

(*g*) After the initial break-through by the armoured formations the first essential is to enable the mechanized, unarmoured, formations to pass through to their support as soon as possible. To this end a further mopping-up and clearance of the route through which the armoured troops have passed may be necessary, and the troops for this purpose must be specially detailed and their position in the order of advance decided. These arrangements will vary according as the position is strongly or weakly held; and the variations within the two extremes will be many.

In an attack against a position of strength and depth the next stage will be the attack outwards from the gap made, carried out in conjunction with the attack against the rear of the enemy's system by troops who have passed through in the wake of the armoured forces. Thus will commence an operation by normal infantry, assisted by infantry tanks and other arms, the object of which is not only to widen the gap but to bring about the complete disintegration, dissolution and occupation of the whole enemy's defensive system on such length of front as may be desired. It is during this operation that I suggest that the troops should act in accordance with the idea put forward in the next section.

(*h*) Although the general plan, and its timings, must be worked out beforehand, it must be emphasized that once the enemy defences are broken through, much more initiative and latitude can, and should be, allowed to lower commanders. Opportunities for exploitation must be taken, and search for these encouraged, so as to keep on striking blow after blow and allow no time for the enemy to recover from the first shock.

6. The Tide Idea

(*a*) Both nature and man, speaking generally, get their way by one of two methods. They use either brute force or the method of disintegration. In the case of nature, these two methods can be illustrated respectively by the tidal wave which breaks down by sheer power all but the stoutest resistance (this is the technique of the armoured attack, the German Blitzkrieg), and by the tide on a calm day which infiltrates, surrounds, disintegrates, and eventually overflows. In the case of man the brute force idea is exemplified by such a battle as Messines (1917) and the disintegration idea by the German advance in March 1918, after the initial break-through had been effected by brute force.

In my opinion it is the tide idea we want to instil into our minds and to develop further.

(*b*) In the tide idea the leading troops do not attack definite resistance. They mask it and infiltrate past it. (Alternatively, if the enemy is in too great strength at the point of contact to allow of the leading troops both masking them and infiltrating past them, the leading troops will do the masking and those *immediately* following will do the infiltration.) It is the troops following behind and definitely detailed for the purpose who isolate the resistance and complete its disintegration; which has been already begun by the mere fact that the leading troops have infiltrated past it. The essential difference between the two ideas is that in the first case the leading echelons of troops fight to overcome resistance, and in the second case they avoid fighting in order to get quickly past such resistance. The actual overcoming of the resistance is no affair of theirs, it is left to succeeding echelons of troops. Therefore frontal attacks are not the business of the leading troops. They mask opposition and pass on. These preliminary maskers press forward further as soon as they are relieved by other troops who have been detailed as maskers from the beginning of the operation.

(*c*) If once this idea can be got firmly into our heads, the necessary tactical organization needed to carry it out will follow. If, however, the idea is not firmly in our heads, the organization will be non-existent and the idea cannot be carried out. Just as in a trench-to-trench attack everything was organized, that is to say, there were 'moppers-up', 'trench blockers', 'consolidators', etc., so must there be organization even for this more fluid type of fighting, i.e. the leading troops must only have one idea, namely to avoid engagement with the enemy which will arrest their advance. These leading troops will be followed by others, whose object is to mask the enemy, or at least to mask such enemy resistance as is encountered. These latter troops relieve those of the leading troops who have been temporarily doing this masking, and allow them to follow the spear-head of the attack. Later will follow the troops whose job is to isolate such resistance as has not already collapsed from the disintegration caused by the tide having passed the resisters and cut them off from the shore.

(*d*) The organization and tactics of the troops who take part in such an operation must be carefully thought out, and organized, beforehand. This is necessary because nothing that is not organized and thought out beforehand will succeed in modern war, whether we like to call the particular type of war we are dealing with at the moment position warfare, open warfare, or loose fighting. These tidal troops are the 'combat teams' of the Germans.

(*e*) To sum up, I feel that constant emphasis laid on the offensive may suggest the slogan 'Stop and Fight'; whereas the better slogan is 'Avoid fighting and Get on', as far as the leading troops in this operation are concerned. I will quote here a short paragraph from

Rudyard Kipling's story of 'The Winged Hats' in *Puck of Pook's Hill*, which I think is apt:

'The Winged Hats fought like wolves—all in a pack.'
'Where they suffered most, there they charged in most hotly.'
'This was hard for the defenders, but it held them from sweeping on into Britain.'

It is the tide idea that the Japanese have adopted so successfully in Malaya.

7. Some Lessons of the last War, and their Application to this one

(*a*) The attack complex during the early stages of the last war caused a useless waste of life in all the armies concerned. For that war showed that large-scale attacks only succeeded where the attacker had:

(i) Great superiority of material.
(ii) Superiority of morale.
(iii) Effective surprise, as well as superiority in numbers and mobility.

For example:

CAMBRAI (20 November 1917). Here we had some hundreds of tanks, the Germans had none, and we effected surprise. We therefore had in our favour elements (i) and (iii). But we had not the mobility to exploit success, otherwise we should have gained a great victory.

21 MARCH 1918. Here the Germans effected surprise, had great superiority in numbers and material, were assisted by fog which blinded our machine guns, achieved a complete initial success, penetrated to a great depth, but lacked mobility to exploit success, without which decisive victory could not be attained.

GERMANS AT AISNE, 1918. They had great superiority in numbers and material, they were opposed by weak formations of tired troops. But they again lacked mobility to exploit success, and so lost a great opportunity of decisive victory.

GAZA III. Here we had everything in our favour, superiority in numbers, material, morale, mobility, air, and we effected surprise. Yet we failed to break through completely, and our large force of cavalry could only cover some 50 miles in 15 days in *direct* advance.

FINAL PALESTINE BATTLE. Here we again had everything in our favour and we achieved all but complete success. Yet note one outstanding lesson in it. Where the enemy morale remained good, complete success was not obtained. That is shown by the fact that those Germans and Austrians who kept together in a compact body and never lost cohesion or morale were never defeated and escaped destruction and capture.

Therefore we can see that big attacks were never completely successful even in those exceptional circumstances when apparently all the determining factors were on the side of the attacker. Even partial success was usually only obtained with very heavy losses. Heavy losses always take the greatest toll of the best troops, so that it is the loss not only of quantity but also of quality which often causes a present victory to be the direct cause of a future defeat.

(*b*) It is now the fashion to say that these lessons are of no value and that while the last war proved the supremacy of the defence over the attack this one has proved the reverse. Is this the case? If so, where has it been proved? When the attacker has everything in his favour, when all the material and moral factors are on his side, of course he wins. So (within limits) he did in the last war. This advantage the Germans have enjoyed in every campaign up to the Russian. In each case the whole requirements of the Blitzkrieg were on parade, from the political advance guard and the fifth column to the tank mass. In each case the opposition was almost negligible, so the Blitzkrieg had an easy victory, and why not?

What about Russia? Here for the first time they met an opponent who was their equal, if not their superior, in numbers, who was reasonably their equal in equipment, and who, most important of all, was their mental equal in the understanding of modern war. In fact, in the Russian Army they met for the first time a High Command who did understand modern war, both on the political and military fronts, and who had prepared themselves for it. It may even be said that from the Russians the Germans had learnt both the use of air-borne troops and the technique of the massed tank attack.

The Russian 'liquidates' those for whom he has no use. Liquidation is a most formidable weapon of war. So the political advance guard was liquidated. So the fifth column was liquidated.

So the unwanted parachutist was liquidated. And the Blitzkrieg was bereft of essential parts. So also the tanks, who were cut off from their unarmoured support, who got into the narrow streets of some country town, who harboured in a wood with insufficient guard, were liquidated. So also all German soldiers who are so unlucky as to lose their way, or become separated from their comrades, or go to sleep....anywhere, at any time, without a proper guard, they are liquidated.

Thus the battle in Russia was different because the Government, the Army, and the ordinary citizens all knew that they were engaged in total war, that in total war everyone is a combatant in some form or another, and that all are in one defence force, fighting together one battle, and fighting it for one end, the destruction of their enemy. We advocate 'Stay Put', which sounds to me like saying 'Get into a funk-hole and stay there'. The Russian stays put until the opportunity occurs to go forth and kill a Boche. Then he stays put no longer, he goes and, with luck, he kills. There are no open towns in the vocabulary of Russia. This is the *mental attitude* that those who would win in total war must cultivate; for, if they do not, they will lose.

Here again is a great lesson for us when we are preparing for the invasion of Britain, or rather preparing the ordinary citizen for the part he has to play in that invasion.

(*c*) In fine, I see no great change to-day in the respective powers of attack and defence provided that both opponents are comparatively equal in mental and material equipment. Modern battles under these circumstances are affairs of months, of years, of alternating offensive and defensive, mobile and static, Blitzkrieg and 'set-piece'. The offensive at the wrong time and under the wrong circumstances will lead to defeat; and with the defensive it is just the same. So it has always been and so it always will be. The defensive which is merely defensive and does not contemplate, and prepare for, the offensive will lose in the end; the offensive continued beyond the point where the gains outweigh the losses is merely preparing itself for destruction by the counter-offensive.

This was the lesson of the last war and it will be the lesson of this one.

Britain is fighting a defensive battle because, for lack of preparation, she must. But she is taking the offensive where she can and is preparing for the great counter-offensive that will be delivered in due course. Thus we are fighting the true defensive-offensive battle, our historic method and our historic role.

We fought this way in the days of the yeomen archers, at Crecy, Agincourt, Poitiers and elsewhere. These were examples of the fire-power defensive followed by the armoured counter-offensive. The knights provided the latter. They were the tactics of Wellington, and Waterloo was a typical example. We fought this way in the last war. Over three and a half years on the defensive, then the counter-offensive and victory in a few months.

So do not let us jump to conclusions, do not let us neglect the teaching of history, do not let us be led away by false ideas of the respective powers of the offensive and of the defensive. To take the offensive before the time is ripe is merely to invite defeat, or at the best to achieve stalemate or partial victory. Let us bide our time, fight the defensive-offensive battle which is our historic role, and, when the time *is* ripe, launch that great counter-offensive which will, as it has many times in the past, lead us on to total victory.

Chapter III

THE GERMAN DOCTRINE OF ARMOURED WAR AND THE DEFENSIVE-OFFENSIVE BATTLE

Part I. SOME ASPECTS OF THE GERMAN DOCTRINE OF ARMOURED WAR

In war fire is everything, the rest is of small account. Napoleon.

March separately, strike together. Clausewitz.

The factors of the 'Blitzkrieg' are *surprise, speed, material superiority,* all given life by *superior fire-power.* F. O. Miksche.[1]

1. Introduction

(*a*) In Chapter II I tried to set out the doctrine of armoured war as it was taught by those who had been studying the subject continuously between the last war and this one. I have done this particularly in order to simplify the explanation of the German doctrine, and to show that in both cases there is the same tactical conception, but amplified and strengthened in the latter case.

The reason for the similarity is obvious. Both sprang from the same root ideas, developed by us in 1917–18. The German High Command accepted the doctrine of armoured war, and gave every encouragement and help to its development. The effect was mental as well as material. In the mental sphere the Germans could think in terms of armoured divisions acting as the main instruments of battle, with all the other arms as ancillary to them, and with full air co-operation.

We, by contrast, had to think in terms of one armoured brigade forming part of one mobile division, acting as an accessory to an unarmoured army. The most we could hope for was one full armoured division in the future; while the situation as regards air

[1] The quotation is from F. O. Miksche's *Blitzkrieg* (Faber and Faber, London, 1941). The author is a Czechoslovak, writing from 'Somewhere in England', and he held high rank in the International Brigade of the Republican Army in Spain.

support was nebulous. We had no really whole-hearted support; our trumpet, unlike that of the Germans, was giving a most uncertain sound. Therefore the mental effect of the respective environments on those who were developing the technique in Germany and England must be taken into account when comparing the results obtained. So also must the material aspect; for, at any rate during the years immediately preceding the war, the Germans had the material to experiment with, while we had hardly any, and practically all that we had was merely of token value. They also had their dress rehearsal in Spain.

(*b*) We had the brains and experience to develop the doctrine of armoured war had we been given the opportunity. There was only one sure way of giving that opportunity, and that was to form a body of people consisting of soldiers who had studied the subject, of airmen to work with them, and of technical experts to assist in developing the material required; and to give this body every facility for developing it. This was not done, no body of thought on this subject was created in the army, and those who had studied the subject were, in the ordinary course of army routine, separated from each other and from the work.

The truth is that our army never accepted the fact that here was a new type of warfare, which in its tactical technique and material requirements was revolutionary, and as such called imperatively for new and special study. Instead, it was treated as an adjunct of the old, and consequently it was not developed either mentally or materially, so that when the war burst upon us not only were we incapable materially of competing with it, but also we were unduly slow in realising its implications.

(*c*) The result of the German study was that their whole method of war was combined into the doctrine of *total war*, military and civil, moral and material. The war on the civil front was prepared politically before hostilities commenced, and the fifth column was organized. When open hostilities began the fifth column merged its activities with those of the military front, and, helped and reinforced by air-borne troops, gave it active military assistance.

Therefore, when we consider the German doctrine of the offensive we must consider total war carried out on both the civil and military fronts at the same time. This is the essence of the Blitzkrieg, and when the civil front does not respond, when the fifth column does not exist, and when the morale of the attacked remains unshaken, then the machine of total war loses some of its essential constituents, and the Blitzkrieg becomes a far less formidable instrument. This is what has happened in Russia, and this is the great, and the most important, lesson we have to learn in the defence of Britain against invasion.

(*d*) We must think of the German doctrine of war as founded on the tactical principle of infiltration; this infiltration being carried out by a machine of many parts, political and economic, military and civil, armoured and unarmoured, in the air and on the ground, all combined into the one machine of total war, all working as part of one whole, on one plan, towards one objective, total victory. In this scheme of war, the tactical principle of infiltration is carried out in the political and economic fields before it is attempted in the military, and the fifth column is just as much part of it as are its aeroplanes and its tanks. Yet just as our own doctrine was founded on the four fundamentals of fire-power, mobility, protection and morale, so is theirs. Just as we, with the limited means at our disposal, catered for the changed factors, so did they, with the greater means at their disposal, re-shape their tactics to suit the weapons they were using. Therefore the doctrines of the Blitzkrieg should *not* be considered as novelties. Study history and you will find that they are not. Rather they should be considered as the application of the eternal fundamentals of war through a new tactical technique.

The factor that has given the opportunity for this new technique is, above all, motorization, for it has enabled us to produce on the battlefield fire-power, mobility and protection in a degree never before attained. But remember always that fire-power is the greatest of the fundamentals, it is the driving force behind manœuvre, the force that makes movement possible, and the only force which can bring victory.[1]

[1] The fact, already mentioned in my extract from Lord Gort's despatches, that we went to war with a tank which had no other armament than a machine gun shows that we neglected the fundamental of fire-power. This in spite of the lesson of the last war that no tank should be without a gun.

The following extracts from *The Times* of 4 February 1942 emphasize the point and, if the comments are correct, indicate that we have also neglected the fundamental of protection:

'Rommel has been greatly assisted throughout his campaign which (*e*) Therefore the military machine which carries out the Blitzkrieg must be considered as consisting of:

(i) The Fifth Column.

(ii) The Air-borne Troops.

(iii) The Air-Component: Reconnaissance, Fighter and Bomber.

(iv) The Armoured Divisions, who form the spear-head of the attack.

(v) The Shock Divisions, who 'mop-up' behind them and open the way for

(vi) The Mechanized, Unarmoured, Divisions who follow the Tanks through the defensive position and exploit success.

(vii) The Normal Divisions.

There are, of course, many other troops who form part of the attacking force, and function at various stages of the operation, such as traffic control, supply, repair, engineers, etc. These, however, concern details, though essential details, of the operation.

2. The Battle (Attack)

'Attack is a Sword, Defence a Shield.'

(*a*) The sword of modern war is the Blitz, with the tank as the thrusting point of the ground attack; the shield is the system of web defence, which slows up and entangles the attack sufficiently to permit of a counter-attack, similarly led by tanks and constituting thus by its strength and suddenness a counter-Blitz.

opened last November by his Mark 4 tanks. These heavy well-armoured monsters, with their 75 mm. guns, have proved a continual thorn in our side. In every way they seem to be the superior striking force, particularly as their equipment—aiming gear, sights, and periscope—is excellent, while the big gun has a range of 3000 yards.'

'In estimating the lessons to be derived from this campaign, one important point for consideration will be whether we have obtained a just balance between light and heavy forces in mechanized warfare. At the moment the outside observer is inclined to believe that we have too large a proportion of the light, so that we are often compelled, for example, to oppose powerfully armed medium tanks with light. The German always ranks the importance of reconnaissance and protection by light forces below that of striking power. German doctrine has often been criticized as disregarding the enemy's reaction, but this policy makes German forces particularly dangerous when the initiative is in their hands.'

(*b*) The tank has three main characteristics:

(1) *It is invulnerable to rifle or machine-gun fire.* But vulnerability is not a fixed point. The technical trend is towards heavier armament and thicker armour, which will make tanks less vulnerable, and give them greater fire-power (cf. p. 48 n.).

(2) *It travels by caterpillar traction.* Therefore it can leave the main road and move across open country. It has thus exceptional elasticity of manœuvre. Some ground, indeed, is tank proof. The tank cannot cross *every* depth of trench or water, nor can it climb every height. But where it is arrested by depth or height, it can often circumvent.

(3) *It is high-speed locomotive fire-power.* Speed combined with fire-power and protection is the characteristic which associates it most closely with the Blitz idea, and any use of it which sacrifices speed sacrifices one of its most distinctive characteristics. The other land arms must conform to it, not it to them. The pace at which it can move is many times greater than that of infantry; even as the aeroplane, which is locomotive fire-power in the air, travels many times faster than the tank. The tank thus at present occupies a pivotal position, midway between the foot soldier and the aeroplane.

The German doctrine of the attack is the same in principle, and the objects aimed at are the same, in tactical conception, as those which I outlined in Chapter II, but they are amplified and strengthened. Their results are obtained by the mass attack of tanks in great depth on a very narrow front, supported by bombers and assisted where possible by air-borne troops. We have a small-scale example of the mass attack reported from North Africa, where seventy of Rommel's tanks attacked ten deep on a front of seven tanks. This method has the following advantages:

(1) The choice of ground suitable to the tank is made far easier than if the attack is delivered on a wide front. Here consideration of the road system both leading up to, and beyond, the enemy's defensive system is important.

(2) The narrow front sets many tanks against few anti-tank weapons. Therefore there is less loss of tanks through anti-tank fire, for the more closely the tanks are bunched together, the smaller the weight of fire from anti-tank guns distributed on a wide front that can be brought against them.

(3) Tanks used as a battering-ram moving at speed require to

use less fire-power to break through than if distributed on a wide front. Therefore the more ammunition will they have in hand for their role of beating the tank counter-attack and creating chaos and destruction in the enemy's back areas.

(4) A mass of tanks covers a small area and requires far less observation by land and air observers than if they are dispersed in small groups. It thus simplifies the direction of the battle.

(5) It is easier for the supporting arms, both air and artillery, to assist the movement of a block of tanks as their supporting fire is not scattered over an area, but falls with concentrated intensity on the narrow section of the front where the tanks are trying to break through.

(6) The great depth makes support from behind constantly available, and ensures that if, on emerging from the enemy's defensive system, a tank counter-attack is met with, there are ample tanks at hand to deal with it.

(7) Whether for meeting a counter-attack or dispersing for attacks against the enemy's rear installations, etc., tanks in this formation are easier for the commander to handle than if they are scattered over a large area of country.

(8) If during an attack the Higher Command wishes to halt it, to withdraw the tanks for use elsewhere, or to alter their direction, it is quite possible to do so with tanks in mass formation, but almost impossible if they are scattered over a wide front.

(*c*) We too had the same ideas in embryo, but the Germans developed three new features:

(i) The provision by dive-bombers of a moving barrage in front of the tanks. The advantage of this over the artillery barrage is that it can be continued to any distance, that it conforms to the speed of the tanks, and not vice versa, and that the observer and the man who fires the projectile are the same person, so that the need for supporting fire and its provision are as nearly as possible simultaneous, and when the tanks eventually split up for their later tasks their covering fire can split up with them.

(ii) The mass attack on a narrow front in great depth. We were thinking on these lines, but had not developed the idea to the same extent.

(iii) The support and assistance given by air-borne troops in the later stages of the attack.

(*d*) The foregoing can be summed up shortly as follows:

The characteristics of the tank determine the form of the tank attack. It converges quickly on the point of attack, the exact point or points being determined not long before zero hour; and it thrusts deeply on a small number of narrow fronts, some of the thrusts not being pressed, but at least one or more yielding the smash through.

The covering bombardment is simultaneous with the attack, being delivered from the air. The purpose of the attack is to find soft spots and to convert the softest into a gap through which the attack can pour. The main attack (or attacks) resembles in its action the punch of a boxer with terrific power behind it, and when successful it has the effect of the knock-out blow. In its initial thrust (its irruption) it gets through any ordinary system of linear defence and creates instant opportunities for itself, and the arms which follow behind it, to take the defenders in flank and rear; and it gets so soon into the heart of the civil population that it creates chaos, setting in motion streams of refugees who terrify each other and paralyse the defence. This chaos is stimulated where possible by fifth column and air-borne troops—a point that we shall consider later in relation to the defence of Britain against invasion.

The tank is the master weapon of the land. But this does not mean that the tank, in conjunction with another master weapon in the sky, is the only thing that counts. The tank only prepares the way for the second stage of the battle, which is the exploitation of the gap created by the tank and of the momentary paralysis created by the two master weapons of land and air. The tank formations (the Panzer units of German nomenclature) are followed closely by other formations which can keep pace with the tanks only because they too are motorized. These consist of mechanized divisions which fight with or from their transport, and of shock divisions, carried forward in lorries which they leave before they go into action. Later, and at a slower pace, follow the normal infantry divisions.

Motorized vehicles of any kind are useless if their petrol gives out, and they are supplied in a variety of ways, by auxiliary tanks (i.e. containers) attached to the vehicles, by petrol-carrying lorries attached to the motorized units, by independent columns, by tanks pushed forward and drawn upon from in front, and sometimes by fuelling from the air.

Let us now follow the action of the shock troops and see how they develop the second stage of the battle. They go into action in 'combat teams' of combined arms, organized and trained to act together. They pass through the gap opened by the tanks, and fan out to right and left.

In trench warfare this meant working along the system and 'mopping up': in Blitz warfare it is known as 'rolling out', which means infiltration into the defender's flank and rear on the lines of least resistance. One combat team helps another. When the defender turns to resist one stroke he is taken in the rear by another. For the combat teams work together, in elastic fashion. There is no fixed limited objective. The maximum of initiative is required from each team. There is no waiting for orders from behind.

Teams vary in size, not only initially, but from time to time as the necessities of the battle dictate. They may be a battalion with tanks and all other necessary supporting arms under command, or even as small as a platoon with a correspondingly smaller amount of support. The action by small elastic groups is, in principle, a development of the infiltration tactics prescribed by Ludendorff for his 1918 offensive. It is the acme of the principle of the decentralized attack—a general plan to which all conform and which lays down the objects for which all strive, but which, once the battle is joined, leaves the initiative to the man on the spot. It is the guiding principle of command to which all should conform, whether that command be exercised on the military or on the civil fronts.

(*e*) The German methods of employing artillery in the attack seem even more revolutionary. A considerable portion of the artillery would appear to be integrated into the framework of the infantry. This does not mean that they are any the less specialists in their own arm, but that a large proportion is used in close tactical support of the infantry both in attack and defence.

I am not prepared to go deeply into the matter here, but it seems that the German artillery are divided roughly, as our machine guns were in the latter part of the last war, into forward guns under command with the combat teams, and rear guns, in larger units, retained under divisional control.

The role of the forward guns appears to be:

(i) To give close support fire against unforeseen targets.

(ii) To protect combat teams during enemy counter-attacks.

(iii) To provide combat teams at any stage of the battle with a firm base, which not only gives them a grip of the ground, but also a jumping off place for a further forward spring.

(iv) To protect the flanks of combat teams.

(v) To provide the anti-tank and anti-aircraft defence of combat teams.

Some of the arguments for this use of artillery appear to be:

(i) To the extent that the air arm can replace artillery more easily in attack than in defence, the attacking role of artillery becomes *relatively* less important than heretofore.

(ii) Anti-tank fire, often at close range over open sights, is a most important task of artillery in defence, because tanks are the main weapon of attack.

(iii) Massed artillery is an easy target for enemy counter-battery bombers. Under air observation a battery that is masked or camouflaged is only hidden till it opens fire.

(iv) The Germans possess a number of assault guns mounted on tracks. The process of development is thus from horse-drawn to motorized, and from motorized to tank artillery. 'The natural evolution seems to be to put as much artillery as possible on tracks—to make the guns in fact "almost tanks".' (Miksche.)

3. Artillery Support of Armoured Forces

As a result of my study of armoured war previous to the present war, and of what has happened during it, I have come to certain conclusions on the question of the artillery support of armoured formations which I insert here, in order that the incidental references to artillery in the preceding section may not be misunderstood.

Put very briefly they are as follows:

(*a*) I consider that field artillery should always be available for the close support of armoured formations in the tactical battle. This is especially important if such formations are acting in a semi-independent, independent or strategical role. In order to do this effectively, the artillery must be able to accompany the armoured troops. Therefore the artillery must be armoured.

As its tactical function is an artillery one, armoured artillery must remain definitely artillery, use artillery methods, and be developed from such artillery as has had experience of working with armoured formations. That is to say, armoured artillery will, as its principal role, be trained for the close support of armoured formations. But, in its tactical conception, it must remain artillery and not become tanks.

(*b*) The following factors affect the issue:

(i) Sustained fire, coupled with the ability to switch this fire quickly and accurately over a wide area, can be provided only by artillery methods.

(ii) Armoured forces may have to operate for a considerable time beyond the supporting distance of other arms, except the air.

(iii) Unarmoured units cannot accompany armoured forces in the area of the armoured tactical battle.

(*c*) The purpose of armoured artillery is to give immediate support with concentrated fire to armoured units and formations in all phases of the tactical battle.

This will entail:

(i) Neutralization of enemy anti-tank weapons and localities.

(ii) Screening the movement of our own forces, probably on a wide frontage and at long range on flank or front.

(iii) Material destruction of strong points, or obstacles, by weight of fire. This may necessitate direct fire at comparatively short range.

(iv) The breaking-up, or disorganization, of enemy counter-attack troops at long range.

(v) Counter-battery work.

(vi) Anti-tank fire with the heavier shell. This is a secondary role and will be employed mainly in self-defence.

(*d*) To fulfil its tactical role, armoured artillery must:

(i) be equipped with a weapon having long range, wide arc of fire, and reasonably high muzzle velocity. It must fire a shell not less in weight than that of field artillery;

(ii) have a cross-country performance equal to that of the arm it is supporting;

(iii) have a speed not less than that of the arm it is supporting;

(iv) be sufficiently armoured to be able to accompany the armoured forces through any enemy-infested areas through which they may penetrate;

(v) have an adequate, and similarly protected, means of ammunition supply;

(vi) be able to protect itself against enemy armoured or unarmoured attack;

(vii) have a flexible organization which will allow it to decentralize sub-units for close support, and to regain centralized control quickly and effectively.

(*e*) Armoured artillery would also be able to act as a strong pivot of manœuvre for an armoured formation engaged in a tank versus tank battle.

(*f*) I therefore consider that regiments of armoured artillery should be created as soon as possible, and specially trained for close co-operation with armoured forces.

(*g*) Armoured artillery must, of course, be able to fulfil a normal artillery role in the zone of the unarmoured battle.

Part II. SOME ASPECTS OF THE DEFENSIVE-OFFENSIVE BATTLE

The whole art of war consists of a well reasoned and extremely circumspect defensive, followed by a rapid and audacious attack. Napoleon.

1. General

(*a*) I have already pointed out that a defensive which is merely static is bound to be defeated in due course. In every war there are periods when one side has to fight on the defensive, and such periods may be long or short, months or years, but they must never be allowed to generate a defensive mentality.

Even in small-scale battles there will be troops not engaged in offensive action who must be disposed defensively. Still more so will this be the case in the great modern battle fought on a front of hundreds of miles. In this case the attacking armies will be fighting a defensive battle on large portions of the front in order that every available resource of men and material may be concentrated at those other points where offensive action is intended. For this very reason the armies of that side which is strategically on the defensive must never allow themselves to become static in thought or action. To do so is to play into the hands of an aggressive enemy by allowing him to hold his defensive fronts with the minimum of strength while concentrating the maximum for offensive operations.

The same applies to the world war of to-day. The allies are, for lack of pre-war preparation, strategically on the defensive. Yet wherever possible they take the offensive as in North Africa or the counter-offensive as in Russia. Even in the cases of Leningrad and Tobruk the defence was never allowed to become static. Counter-offensives, large and small, have characterized the defence of both places. For although the object of defending a place or area is to prevent the enemy occupying it, such defence must also always aim at containing as many enemy troops as possible. In this way the enemy's main offensives are hampered and our own operations assisted.

(*b*) Therefore all defensive operations must be of a defensive-offensive nature; and all those participating in such operations must be imbued with the most aggressive determination to make their defensive as offensive as possible. As I have said before, it is the mental attitude towards anything which is the most important. If we have the right idea we shall succeed in the end, however many mistakes we may make on the way. If we have the wrong idea we shall fail.

The right idea, as regards the defensive, is this, that while for a time the defensive may be imposed upon us we must never become obsessed by defensive thought, we must never become static, we must never forget that our job is to kill the enemy and not to make ourselves safe, except in so far as making ourselves safe helps us to kill better, and that whatever part we may be playing, that part can have only one ultimate object, namely the taking of the offensive and the destruction of the enemy.

(*c*) Formerly, and in fact till quite recently, teaching in the British Army stressed the importance of anti-tank obstacles, natural ones like rivers, canals, etc., or artificial ones like those dug in parts of England in the summer of 1940. But it was not stressed sufficiently that no obstacle is an obstacle unless there are sufficient men and weapons to guard it at all points. Unless men and weapons are available in very large numbers the attempt to guard such obstacles will usually result in the men and weapons available being 'strung out' in a linear defence on a long front. Such a form of defence will be able to offer no resistance to the massed armoured attack, it will not even delay it to any appreciable extent.

Again, unless a zone of some depth *in front* of each obstacle is held, the enemy's unarmoured troops will be able to occupy the far side of it, prepare a concentrated attack at some point or points on it, make a bridgehead over it, prepare a crossing, and launch the armoured attack.

The purpose of holding a zone in front of the obstacle is to prevent the unarmoured troops penetrating to the obstacle. It may be impossible to prevent the armoured troops doing so, but they cannot cross without the help of the unarmoured, and therefore the forward defended zone has got to be captured before a crossing over the obstacle can be effected. Thereby the armoured attack will either be delayed, while it awaits the reduction of the forward defences by other troops, or will become separated from its unarmoured troops and gain no assistance from them in its subsequent operations. If the armoured troops determine to reduce the defences themselves, they will be delayed and also caused to expend ammunition which they will urgently require later.

Therefore a combination of obstacles, and defences in front of them, will tend to cause delay at a time when the direction of the armoured attack has manifested itself, and at least will cause the enemy tanks to expose themselves to the fire of the defence for a longer period than if they had been able to push straight through without delay.

The effect of separating the armoured troops from the unarmoured is to deprive the former of support in their initial attack, and of subsequent help during exploitation. Experience has shown, however, that no anti-tank obstacle, natural or artificial, has been able to hold up a determined attack for any length of time. In fact the only one, and that a sea obstacle, which has so succeeded is the Straits of Dover. So reliance must not be placed on such obstacles except in so far as they fit in with a complete system of defence in depth, delaying the armoured attack in one place, canalizing it in another, enabling fewer anti-tank weapons to be used in one place and more in another, the obstacles thus assisting in the building up of the defensive system, but not dictating its position.

Lastly, we must remember that in all our defensive measures we have in view the delivery of a counter-offensive by our own armoured forces, supported by such mechanized troops as we possess, which, if successful, will be exploited with the whole of our resources. Therefore from the beginning we must dispose our counter-attack troops, so as to facilitate the initiation of such counter-offensive as soon as the opportunity for launching it presents itself.

2. Principles of Defence against the German methods of attack (based on Miksche's *Blitzkrieg*)

(*a*) As we have already seen, the principle of the German attack is the breaking-through of the defensive system by the mass attack of tanks on a narrow front, and the disintegration and dissolution of the system by the combat teams that follow in the wake of the armoured mass. Thus we have two main problems of defence, that against the armoured attack and that against the infiltration of the combat teams.

The task of holding the defensive system falls to the unarmoured troops, the normal infantry divisions.

The armoured troops, both armoured divisions and infantry tank brigades, must be concentrated under central control in rear of the defensive system, since their role is the counter-attack against such enemy forces, especially armoured ones, as succeed in breaking-through that system.

(*b*) Tanks themselves can achieve no permanent success. The attacker can only win a real victory if, in addition to the rupture by the tank mass, his combat teams can so clear the way through the gap created that the main body, the normal infantry divisions, can be passed through the defensive system. Therefore the first object of the defence is to separate the enemy's armoured forces from his unarmoured, and to prevent the latter, together with the tank supply services, from following the former.

Hence the first requisite of a defensive system is to be able to offer strong resistance *after* the enemy tanks have passed through it, and so cut off the hostile mechanized formations from their supporting troops and from their supplies of fuel and ammunition. The second requisite is to gain time, to delay, hamper, hold up, and even if they cannot stop the armoured attack, to cause disorganization and loss. If they can compel the enemy tanks to fight and expend ammunition, it will greatly advantage our own tanks when they meet them in the counter-attack.

(*c*) To achieve these objects the main basis of the defensive system must be a network of islands of resistance. Weapons, located in anti-tank localities, will cover anti-tank obstacles and also the ground between the islands of resistance, which are themselves anti-tank localities. These weapons not only fire against the tanks and their supporting troops but prevent infantry and engineers from getting to the obstacles and making them passable for tanks. The system must be based on the greatest depth that our resources in man-power and weapons make possible. While depth is essential, it must not be carried to an extent which renders the system weak everywhere; it must be deep, yet strong, throughout.

It must be designed to disintegrate the attacking force. It is like a filter, it lets some parts of that force through, but at a cost, and it retains others—retains them for destruction.

The islands of resistance must be so organized that they can fight independently even when surrounded by hostile forces. The attacker is thus forced to fight a number of separate battles, his units and his supporting fire become dispersed, so that both are less effective. The invisibility of the defence is important, but as far as the islands of resistance are concerned this must not become a fetish. It is true that a weapon must be invisible if it is to function effectively and continuously; for if, the moment it opens fire, its position becomes obvious, it will soon be silenced. But islands of resistance will frequently be best formed from towns, villages, farms, woods, etc. Built-up areas, though certainly they will attract fire, give cover and concealment for the defenders and they are easily made tank-proof. In 1917 and 1918 the Germans made the most effective use of fortified villages, and during the Civil War in Spain fortified towns and villages formed strong and effective centres of resistance. Towns and large villages, defended by determined troops, are very difficult to take, can be made death-traps for tanks, which cannot dig men out of whole streets, are good bases from which combat teams issue for counter-offensives, and form pivots of manœuvre for counter-attacks on a larger scale.

Against the tank irruption, that is the advance of the armoured mass, the unarmoured defenders will draw themselves within their islands of resistance, their anti-tank weapons remaining in action against the tanks. When, however, the unarmoured troops appear, the defence must become aggressive to the greatest possible extent. Every island of resistance must have its combat team, so that the attack on the enemy will not only comprise fire from the island but also well-directed counter-attacks by the combat teams against the flank and rear of the attackers.

Thus the static fire of the islands and the mobile fire of the combat teams will be combined. The defence becomes an active one, every opening given by the enemy is taken advantage of, and the old principle of static and mobile fire-power in combination is used to the best effect. In such a defence an intimate knowledge of the locality will be of the utmost advantage. The ideal defender of such islands of resistance is a people's army of Home Guards who, from the very fact that it is their home, know intimately the place itself and the surrounding country.

(*d*) I will not go into details of the organization of the defensive system but merely stress the salient points.

(i) The garrisons of islands of resistance may vary in size, according to the extent of the island and its tactical requirements, from a company of infantry to a brigade. All garrisons have their close-support weapons, machine guns, anti-tank guns, mortars, while the larger ones have artillery in addition. Thus each is really a self-contained garrison. Within each garrison are organized one or more combat teams of shock troops, whose role is offensive action against enemy units infiltrating in close proximity to the island. Thus each island is organized for the defensive-offensive battle.

(ii) Light machine guns are mainly for the protection of the island itself, medium for the protection of neighbouring islands and to provide covering fire for offensive action. Personally, as an old machine-gunner, I suggest that we hold to the system which we worked out after much trial and error in the last war, namely that the medium machine guns are divided into forward guns under command in the larger islands of resistance, and rear guns retained under divisional control. Thus the divisional commander will always have a reserve of artillery and machine guns at his disposal.

(iii) In some cases it may be better for islands to let the tanks pass and reserve their fire for surprise effect against the enemy unarmoured troops who are following the tanks.

(iv) As in the attack, so in the defence, the doctrine of decentralization and cultivation of the initiative of the lower commanders holds good.

The higher commander makes a general plan for the defence and co-ordinates the efforts of the defenders. He must never, however, allow his command to become so detailed as to result in rigidity and lack of initiative among subordinate commanders. Success in the defence, as in the attack, is attained by the full use of the initiative and individual powers of every man and every weapon. This doctrine necessitates the highest level of morale throughout all ranks, each individual must be prepared to go on fighting, and fighting aggressively, even when all seems lost within the area of which he himself knows anything. Of this the Russians in their stern trial of 1941 gave many a brilliant example.

(v) When built-up areas are held, fire-guards, firemen, appliances and water must be available.

(vi) The Germans make great use of mines and consider them to be in many ways better, and more economical in transport, than would be a sufficient amount of artillery ammunition to produce the same stopping effect against tanks.

(vii) All islands of resistance must have sufficient stocks of supplies of all kinds to enable them to continue fighting for several days after they have been surrounded.

3. The special case of Defensive Action by Armoured Forces

It may be necessary for an armoured division carrying out an independent or strategical offensive role to act on the offensive-defensive or even, for a time, definitely on the defensive.

Two of the ways in which it can do this are as follows:

(*a*) It can seize an area, or areas, vital to the enemy, which that enemy must regain: establish its unarmoured support troops on this area or areas, and withdraw its armoured troops into such a position as will enable them to act offensively against the enemy should he attack those left in position. Armoured artillery with the unarmoured troops would be of great value in such a case.

(*b*) It can establish what is called a 'tank-infested area'. That is to say, having chosen a suitable piece of country, it will allot areas to tank units in which all routes are picketed by tanks, a central reserve being retained with which to strike at favourable targets which present themselves. In effect there is presented to the enemy the problem of movement through an area literally infested with tanks, which lie up in concealed positions, take action when he appears, move to fresh positions, when threatened, and repeat the process as often as necessary.

4. Summary

(*a*) The defence consists, as it always has consisted, of two main parts:

(i) What is aptly termed in *Blitzkrieg* the web defence—a system of defence which aims at netting, weakening, slowing up, and eventually immobilizing, the attacker.

(ii) The large-scale counter-attack, or counter-Blitz, which aims at his defeat and destruction.

Therefore it is founded on Napoleon's dictum, for it depends for its success on the 'well reasoned and extremely circumspect defensive', without which 'the rapid and audacious attack' will not succeed.

(*b*) Miksche divides up the whole area to be defended into zones—outpost, main (or filter), guerrilla, and reserve.

The names of the zones indicate their roles. But the motif which runs through the whole idea is flexibility, initiative, aggression. The first job of every man and first object of every arrangement is to kill the enemy. To that all else is subordinate.

The factors in war that have made this mental attitude of supreme importance are the tank, linked closely with the air arm and with mechanized unarmoured troops. These factors have brought speed into every aspect of war and they have, because of that speed, made instant action imperative; this has given a new importance to the initiative of the individual.

(*c*) The defensive system, as we have seen, will be manned by the unarmoured troops, the normal infantry divisions, while the counter-offensive forces must be principally armoured troops. They must deal with the break-through of the tank mass. If they can defeat that, the rest is easy. Deprived of his armoured spear-head the enemy is helpless in face of such armoured formations as the defence has left. At present the only real answer to tanks is more and better tanks; but already the special heavily armed and thickly armoured anti-tank aeroplane has appeared over the battle. Here is a development that those who follow the Four Fundamentals will watch with interest.

The defender will have, almost certainly, less tanks than the attacker at the beginning of the operation. His job is to destroy, disperse, and disorganize, as much as possible, the enemy tank forces, so that when they do emerge from the defensive system his smaller number can deal with them effectively. The counter-attack is more likely to be successful if it can be launched while the enemy's armoured forces are still tangled in the web of defence than if they are allowed to emerge, and reorganize, in the open country beyond.

(*d*) The enemy will have made every effort to gain air superiority over the area he has chosen for his attack. In fact, if he has not gained such superiority, his attack will almost certainly fail. Therefore, before the counter-attack can be initiated the defender must make every endeavour to regain air superiority over that area in which the counter-attack is to be launched. For this reason the strategic reserve of a modern army must always include a strong air component.

(*e*) In defensive operations it will always be a race in time between the advance of the attacker and the assembly in the right place and at the right time of the counter-attack forces. That is why it is so important that the defensive system should delay the Blitz attack for as long as possible. For the attacker, having the initiative and having chosen the point at which he intends to make the rupture, will always be able, in boxing language, to beat the defender to the punch. Hence every minute that the attacker can be delayed on his way through the defensive system is a minute added to the time which the defender has in which to arrange and deliver his counter-stroke. The method of delivering this counter-stroke invites careful study. The method will change with the ever-varying factors, but the technique of its delivery as discussed in *Blitzkrieg* is worth careful consideration.

(*f*) What if the enemy leaps over the defensive system with air-borne divisions? Here comes in the importance of the defence of the whole hinterland of the defender which in the case of Britain must be the whole country.

This means the organization into islands of resistance of all towns, large and small, of all important villages, in short the creation of a system of web defence throughout the length and breadth of the land. These islands of resistance must not only resist attack by the air-borne troops, they must be the harbours from which issue the counter-attack combat teams who attack the air-borne troops as soon as they land, who harass them wherever they appear, who prevent one body of them joining up with another, and who by harrying them incessantly, day and night, immobilize them until they can be destroyed by a counter-attack on a larger scale.

Thus defence against air-borne troops is based on exactly the same principles as any other defence. First, you have the static defence of the islands, which prevents the enemy gaining possession of important strategic and other localities. Secondly, you have the counter-offensive combat teams, who issue from the islands and who, together with the patrols and 'guerrilla bands' of the Home Guard, on ceaseless watch throughout the land, harass and immobilize the enemy. Thirdly, you have the counter-attack formations from the general reserve, which finish him off if this has not been done already. *In preparation for air-borne attack the defence of aerodromes is of vital importance. To lose aerodromes is to lose the first phase of the battle; to lose many is to court final defeat.*

(*g*) In conclusion it may be said that a study of this most modern system of defence shows once again that the great principles of war are eternal, and that if we have studied the history of war well and truly, we shall see how the application of these principles is varied from time to time to meet the ever-changing factors.

To-day, as from time immemorial, there are only two methods of dealing with an enemy's attack: defence and counter-attack. Defence repels the enemy by fire, counter-attack defeats him by fire and movement. Attack is a sword, defence a shield. Only the sway of battle can dictate which method of fighting rules at any given moment and at any given place. A real tactician fights always with both shield and sword, alternatively or simultaneously. A great commander, whether he be fighting one campaign or the many campaigns of a world war, whether he is forced to act on the strategic defensive for months, or for years, will fight that war on the principles of the defensive-offensive, will take the tactical offensive whenever the circumstances are appropriate, and will always prepare for the day when he can pass to the strategic offensive, and launch that counter-offensive which alone can lead to total victory.

Chapter IV

THE DEFENCE OF BRITAIN AND THE CIVIL FRONT

Part I. THE DEFENCE OF BRITAIN

And if a kingdom be divided against itself, that kingdom cannot stand. And if a house be divided against itself, that house cannot stand. And if Satan rise up against himself, and be divided, he cannot stand, but hath an end.

St Mark iii. 24–26.

1. Introduction

(*a*) The defence of Britain is, relatively, the most important thing in the whole War. For she, as I have said (p. 9 above), is the headquarters of the anti-Nazi front, the fortress aerodrome which forms the base for air and sea attack against Germany and the acknowledged head and centre of the resistance to the Axis Powers. Others may fail and the fight can still go on. Let Britain fail to preserve herself, and Hitler dominates the world. What a stake for him to play for! Remember that the so-called impossible has never daunted him.

(*b*) There are still many people who argue that an invasion of Britain cannot take place. I believe that it can, and in sufficient force to make the task of defeating it a most formidable one. I believe that the idea that complete air supremacy is necessary over the country to be attacked is out of date, and that the Germans will no longer adhere exactly to the procedure which they adopted in 1940. I believe that reasonable air superiority at certain places, and at certain times, is enough to permit of very large numbers of air-borne troops being landed in this country. I also believe that the same applies to the sea. That is to say, reasonable air and surface superiority in certain areas would enable considerable forces to be landed on our shores. The Germans have now far more experience than they had in 1940, and with this experience behind them I think that their technique will be different from what it was then. Therefore, do not let a false sense of security lull us into a state of complacency, rather let us realize the danger and leave nothing undone to prepare ourselves to meet it. For meet it we can, and with success, provided we are fully prepared, mentally, physically and materially, on both the civil and military fronts. The Germans must realize by now that they cannot win the war unless they can destroy Britain, and that the quickest and most effective way of destroying Britain is here in this island, which alone, and for so long, barred their road to world domination. That they will come I am sure, whether it be as victors from another theatre of the war or as the gamblers making their last throw.[1]

Let us also be in no doubt that when they come they will strike with all their might and with every resource of men and material that they can put into the attack. In our preparation for that attack we must not forget that gas may play an important part in it. It will be the greatest Blitzkrieg of them all. So let us have no illusions and let us prepare, let us one and all gird ourselves for the battle, and leave nothing to chance.

2. The Methods of Attack

The attack on Britain will take three main forms:

(i) *The Air Attack on our morale and on our material resources.*

This is the battle that the Civil Defence Organization, with a certain amount of assistance from the fighting Services, has been engaged in since September 1940.

If the 'Battle of Britain'—that is, the daylight attack on this country—had succeeded, it would probably have destroyed at one and the same time both the physical power to resist and the will to win. It would have been followed by air-borne, and most likely by sea-borne, invasion. But it was defeated. Therefore, the lengthy and comparatively ineffective method of high-altitude night bombing was resorted to. It was not so effective as might have been expected, but it is foolish to disguise that its effects were serious, and that they may be still more serious in the future.

[1] I should be glad, of course, to be wrong here. But right or wrong I urge this. If developments on the continent make invasion less probable, this should provoke not a relaxation but an intensification of the preparations by the Home Guard and the components of the Civil Front. For the possibility of troops now in Britain being released for activities overseas to take advantage of the enemy's weakening will depend in no small part on the confidence which the Government can place on the ability of the Home Guard and the Civil Front to fulfil their maximum role in the defence of Britain.

(ii) *Air-borne Invasion*

This would almost certainly be preceded, and accompanied, by heavy bombing attacks designed to dislocate communications and to confront us with the problem of dealing with a mass of bombed-out and homeless people at the very time when all our energy should be devoted to resisting the invader.

Therefore, complete arrangements must exist to deal with this problem, and thus to prevent loss of morale and consequent demoralization from impairing the fighting spirit of the population. On this question I shall have more to say.

The attack may take the form of a vast raid: the intention being to land large bodies of troops at a number of different places, to wreck ports, communications, factories and other important war installations, and thus to do as much damage as possible to our war effort. Even if the raiders were eventually destroyed, they might well succeed in first inflicting widespread damage and dislocation. On the other hand, it might be the preliminary to a sea-borne attack. In any case it is certain that the latter will be prepared, so that if the raiders achieve a considerable measure of success they can be supported from the sea without delay.

(iii) *Sea-borne Invasion*

This, if it were the real thing and not merely a series of raids, would mean that the Blitzkrieg against Britain was fully launched. We have seen that to be complete it should be preceded by its political advance-guard and that its fifth column should be mobilized and at work spreading rumours, fear and demoralization; initiating false orders and ready to render active assistance to the air-borne troops when they arrive. We may feel confident that under our Prime Minister's great leadership we are politically sound, and we may leave the fifth column to the Home Secretary, the Police and Regulation 18 (*b*).

But here is a most important point in our preparation. Ignorance of what is happening, or is likely to happen, of what one ought to do, and how one ought to behave, the inevitable rumours and false news, and the fear that comes from not having a definite job to do, will create a large number of unintentional fifth columnists. If we do not take care, these people may, quite unwittingly, become almost as dangerous as the real article. Ignorance and lack of a job are two things which, in a crisis, are most damaging to morale. In our preparation for invasion we must insure that people are not ignorant and that all know the part they have to play.

The preparation of the civil population I discuss below (p. 106), for in total war it is just as important as military preparation. But, assuming that we have deprived the Blitzkrieg of two of its most important constituents, treachery and confusion, and that thereby we have liquidated, even if not quite as ruthlessly or efficiently as in Russia, its political advance-guard and its fifth column, the Battle for Britain becomes an operation of war, not merely of war on the military front, but of total war; of war on both the military and civil fronts, in which all must play their part who have the mental and physical strength to do so. Wherever the Blitzkrieg has succeeded, it has found the Kingdoms and the Houses divided against themselves, and Satan united. In Russia for the first time it met a country united in Kingdom and in House, and prepared for total war. Therefore, it failed for the first time, and for the first time there appeared signs that Satan was not as four square to himself and to the rest of the world as heretofore. Let us make certain that when Satan comes to Britain he finds that the Kingdoms and the Houses, the military effort and the civil effort, the authorities and the people, and the people among themselves, are not only not divided, but by *training and preparation* are so welded together into one solid front of total war that Satan's impact against that front will cause still further division among the powers of hell.

3. The Methods of Defence

(*a*) When the invasion comes, Britain will be strategically on the defensive. As we have already seen, the strategic defensive must be based on the principle of the defensive-offensive, which consists of two main parts, the static defensive by the troops holding the islands of resistance which constitute the system of web defence, and the counter-attack of troops whose success depends on that of those who operate the static defence.

We have also seen that even the static defence must be of an aggressive and offensive nature and that counter-attack teams must form part of the garrison of every island of resistance.

(*b*) I do not intend to attempt to deal with the counter-attack portion of the defence. It will consist of armoured, mechanized, and motorized troops of the Field Army and is entirely a matter for the Higher Military Command. The static defensive is different, for it, to be successful, must be an intimate amalgamation of the civil and military fronts.

(*c*) For this part of our defence Britain must be organized throughout into one vast system of web defence. Towns large and small, villages, aerodromes, factories, etc., must be organized as islands of resistance. They must be able to place themselves in a state of defence at any moment. Remember that one of the essential techniques of the Blitzkrieg is surprise. Remember that the Germans landed 5000 fully equipped troops at Rotterdam airport in half an hour. If once these air-borne troops are allowed to establish themselves, to seize aerodromes, to occupy towns which are nodal points of communication, and to dominate areas of the country, it will not be long before the avalanche of modern war is upon us, and the Blitzkrieg is in our midst. Then, if this has happened in a number of places all over the country, the counter-attack forces, which cannot be everywhere, will have so many targets to attack, so many situations requiring urgent attention, that they may well find their strength dispersed and the situation out of hand in many places. It would be like what happened in the early heavy incendiary attacks on our cities, before fire-watching was understood and organized. Fires started in dozens of places at the same time, the Fire Brigade could not be everywhere, the fires got out of hand, and even if there had been the appliances, water in sufficient quantity was not available.

Thus the fires became conflagrations, whole areas of towns were destroyed, and the fire brigades had to give up fighting the fires themselves, and had to try merely to prevent them spreading beyond certain places and to allow them to burn themselves out elsewhere. So the necessity of the Fire Guard became apparent, and finally it was realized that Fire Brigades are the counter-attack troops, while the Fire Guards are the defensive system, and that, as in the defence against ground attack, so in the defence against incendiary attack, unless the defensive system does its job the counter-attack troops cannot do theirs.

Who are the fire guards of our national defence against invasion? Who are the people who hold the national system of defence and thus enable the counter-attack forces to fulfil their role effectively? The Home Guard. They are the people on whom the defence of

Britain in the first instance depends, they are the people who must be ready for instant action, they are the people who must put out the first fires of the German attack, and prevent them from becoming conflagrations, and thereby so reduce the number of 'incidents' that the counter-attack troops can cope with them successfully. In dealing with the incendiary bomb speed is essential. A few minutes' delay and the fire has got going and has grown quite out of hand. It is the same with the action of the Home Guard against the air-borne troops. Dealt with at once it is comparatively easy, left a few minutes it may be beyond the power of local forces to subdue. Therefore, the Home Guard, like the Fire Guard, must be on the spot, know every inch of its area and be at all times ready for instant action.

(*d*) The Home Guard was created and organized for the express purpose of being the static defence of Britain. Whether at the time of its creation those who created it had any clear idea of how it would carry out that defence I do not know. But there must have been a fairly good idea somewhere; for its original name, Local Defence Volunteers, implies precisely the role which I have suggested and for which I understand it is organized and trained. I have no intention of going into details of organization and training. The only things I will touch on are these:

(i) The Home Guard, because it is responsible for the defensive side of the defensive-offensive battle, is the *first line* defence of Britain, for unless it does its job properly the counter-attack troops cannot launch their 'rapid and audacious attack' effectively, and the battle of Britain will be lost in advance. The Home Guard will, of course, be helped in its task by all the static troops in the country, in schools, depots, training centres, etc., but it is the backbone of the home defence, and the success or failure of that defence will rest with it. Every man of the Home Guard can say to himself, and say rightly, 'It all depends on me'.

(ii) The defensive system which the Home Guard holds must be a system of islands of resistance throughout the length and breadth of Britain. There must be no attempt to hold 'lines of defence'; for lines in a defence of this nature are of no avail against the German methods, they are only waste of men, labour and time. The Home Guard will hold their fortresses, big and little, throughout the land. From these fortresses will issue the combat teams and the guerrilla bands, which will harass the enemy by day and night and destroy not only his bodies but his nerves. There is little doubt that the work of the guerrilla bands in Russia has made a contribution to the Russian resistance so great that without it the German failure could never have been brought about.

(iii) There is no need for every man in the Home Guard to be young and active and super-fit. The large majority will be required only for the purely static defence of the islands of resistance. As long as they can use their weapons efficiently, that is all which will be required. By carrying out that part of the defence which is purely static they release the young and active for the combat teams, the patrols, the stalkers and the guerrilla bands. Therefore, the more men there are who can use their weapons statically, the more younger men will be released for active operations. Remember that every defence, however small, has its two parts, static and active, and that the active is based on the static. Age does not enter into it, the only question to be asked is: 'Can this man use a weapon efficiently in a static defensive role?' If the answer is 'Yes', then he has a most useful part to play in the Home Guard.

(iv) This war is a war of nerves. It will be won by the side that destroys the morale, the will to resist, of the other side. It is a war which turns every citizen, man or woman, old or young, into a military objective. Therefore, it requires a different point of view towards war from that which was good enough for civilians in the past. No longer can the civilian say: 'You fight and we will help, but you must not expect us to take an active part in the fighting ourselves.'

In this war the civilian is in it just as much as the soldier, he must expect to have to hit back, to help actively in the battle. This is a total war and it calls for the total defence of every town and village by its inhabitants. Not only must they hit back in order to defend, but also to make certain of keeping their hearts up, of retaining a good morale, which, without a definite job to do and a definite part to play, they will not succeed in doing. But it is imperative that they should do so; for now more than ever good morale is the master key to victory.

(v) I have no intention of entering into the controversy as to whether civilians should fight, and all I will say in that respect is that numbers of untrained and unorganized civilians joining in the battle would probably be far more of a nuisance than an assistance. But there are many ways short of actual armed intervention in which unarmed civilians, both men, women and young persons, can assist in the defence of Britain, if by previous preparation, organization and training they have been rendered capable of doing so. That is one of the things that the Defence Committees which I describe in Chapter v should take in hand.

The ideal would be that every man in the country is trained to arms and organized for military action. Then, whatever his normal task, when the enemy is at the gate, he ceases to be a civilian and becomes a soldier. This is, in our case, impossible and could only be achieved by adhering in peace time to the ideal that the first duty of every citizen is to fit himself to defend his country with arms in his hands. This has not been our ideal in the past, and therefore cannot be implemented in the present, but it is to be hoped that the lessons of this war will cause it to become our ideal in the future.

But it does seem to me that all able-bodied men of the Civil Defence Services should be trained to arms. Even if, at the moment, it is not feasible to arm them fully, they should be trained, so that they are capable of using such weapons as will become available later. They should form a section of the Home Guard. Their primary work as heretofore will be civil defence, but when the enemy is upon them and civil defence work is no longer possible, they should be able immediately to cease their work as firemen or A.R.P. workers, take up their weapons, and assist the Home Guard to defeat the enemy.

(vi) The Home Guard is a civilian force. Everything it does will be done amongst, and in the closest contact with, civilians. The success of every step taken to resist, and defeat, invasion, will depend on the closest co-operation with civilians. They and the civilian population must plan together, organize together, and resist together. If this spirit animates all concerned, if secrecy is a thing of the past, if all know the part they have to play, and if initiative and a readiness to take responsibility and act without waiting for orders, or reading a regulation, are encouraged, then there will be no loss of morale and no failure on the day of trial. So I will sum up by quoting from two books which all should read. First, from that of Mr John Langdon-Davies, *Home Guard Warfare*:

in which he advocates regular local defence meetings, and writes as follows:

> 'A regular town meeting or village meeting to which every adult inhabitant goes to learn how his locality is to be defended and what part he or she is to take in that plan of defence....So let's have our meetings to discuss the problem of Defence... and let us get rid of the watertight compartment idea of the job before us—Home Guard, A.R.P., N.F.S., and the rest must dovetail into one another, and there must be one unit and one unit only—the [Town or] Village as a whole. The Total [Town or] Village organized for Total War.'

Secondly, from that of Mr Hugh Slater, *Home Guard for Victory*, in which he says:

> 'The Home Guard is itself half civilian. It is a people's army. How useful it can be in the military sense depends wholly on the extent to which it reflects the needs, the desires, and the aspirations of the ordinary British people. Its purpose is a democratic one—to win the war against Fascism. The Home Guard must, therefore, be thoroughly permeated with democratic ideas, methods and attitudes.
>
> 'In contemporary warfare there can be no sharp distinction between the army and the people....
>
> 'One of the most important lessons of the campaign in Flanders is that it is absolutely impossible for the army to work efficiently without civilian co-operation. The hundreds and thousands of French civilian refugees could only be regarded by the armed forces as a nuisance. There was the illogical situation of an army fighting for the interests of a population whom that army had to regard as a military menace.'

He goes on to point out that if the Home Guard establish the closest co-operation with the civilians in their areas they can become a most valuable bridge between the general public and the army and thus increase that integration of the army and the people which is so essential if our war potential is to be developed to the full.

Thus, both these authors stress the motif which runs through these Lectures, the complete integration of the civil and military fronts and consequent formation of one united front for total war.

Part II. THE CIVIL FRONT

Referring to Britain's part in past wars *Time and Tide* says:

In all those wars there has come a time when the enemy has worn himself out by conquering, a process in which *their* [the Allies'] defeat was education and *his* victory delusion.

But the last throw of invasion we cannot overlook. Even if the entry of the United States makes a victorious invasion little more than another stage in the process of Germany's defeat through barren successes, we cannot ignore the fact that if invasion by any chance succeeded—and against resources which are still far from exhausted and would be used to the full, our defence might conceivably be temporarily broken—then whoever might win the war, we should have lost it.

Extract from *Time and Tide*, 27 December 1941.

1. The Civil Defence Organization

(*a*) The Central Department mainly responsible for Civil Defence is the Ministry of Home Security, but other Departments are vitally concerned in certain problems, e.g. the Ministries of Health, Food, Transport, etc., and in order to secure consultation, co-ordination and collective decisions over the whole field, there exists what is called a Civil Defence Committee of the Cabinet. England, Scotland and Wales are divided into twelve Regions. The Ministry of Home Security is in direct communication with the Regions, collects all intelligence from them, and is a channel of communication between the Government and the Regional Commissioners.

(*b*) Regional Commissioners are charged by Royal Warrant with the responsibility of ensuring that the Regional plans of the Departments responsible for Civil Defence and those of all Local Authorities are properly co-ordinated. If communication with the Central Government becomes very difficult or impossible, the Regional Commissioner is empowered to act on behalf of the Government.

The staff of Regional Headquarters consists of the Regional Commissioner, one, or in some cases two, Deputy Regional Commissioners; a Principal Officer, who is a Senior Civil Servant, with a Deputy Principal Officer and a considerable staff; Regional Officers, whose duty it is to deal with all A.R.P. matters throughout the Region; a Chief Regional Fire Officer and his staff; an

Inspector of Constabulary and a Police Staff Officer; a number of specialist officers and Military and R.A.F. Liaison Officers, whose duty it is to keep close liaison between Regional Headquarters and the Military and R.A.F. Authorities. In addition, practically all Ministries and Departments of Government are represented in the Regional Organization, and maintain contact with Regional Headquarters through their Senior Regional Officers, or through Liaison Officers. There is also a Treasury Officer attached to Regional Headquarters.

Thus, the Civil Defence Organization covers almost every aspect of the life of the nation and practically all these aspects are represented at Regional Headquarters.

(*c*) For purposes of Civil Defence, Regions are divided into A.R.P. Scheme-making Authorities. Each Scheme-making Authority has a Controller and a Deputy Controller; a Control Centre which is in communication with Regional Headquarters and with a varying number of Sub-Control Centres, which again have Report Centres in communication with them. So the whole Region has a system of Control and Report Centres situated all over it and a ramification of communications throughout its whole area. Therefore, the Civil Defence Organization has a network of communications stretching from Home Security down to the local Report Centres, and furthermore, through the Regional Commissioner and his departmental Liaison Officers, has a system by which government can be carried on throughout the country even if there were a temporary breakdown of control at the centre.

Each Scheme-making Authority has at its disposal not only the system of communications already described, but is in close touch with Police, Fire, Medical, Engineering, Food and other Services.

2. Civil Defence Resources at the disposal of the Military

(*a*) It will be seen from the foregoing that the Civil Defence Organization and the Local Authorities have many resources which could be placed at the disposal of, and be of great value to, Military Commanders. To enable them to obtain this assistance, it is of the utmost importance that Military Commanders should make themselves acquainted with the personalities and machinery of the Civil Defence Organization in any area in which they are proposing to operate. The three most useful officials to get in touch with are the A.R.P. Controller, the Chief Constable, and the Clerk of the County Council or other Local Authority. In many cases the same man may combine two of these appointments. Any one of these three persons can give the necessary information as to what individuals, or bodies, can give the Military the best assistance for any particular purpose. It is necessary to find out this, as the organization for close co-operation between the Military and Civil Authorities is not exactly the same in all Regions. For example, in the South-West Region we have a complete system of Defence Committees, on which sit Military representatives, at every level of Local Government, from Region to Parish.

(*b*) County Councils have a staff of qualified road engineers supported by a trained and experienced labour force, fully acquainted with the localities in which they live. They are equipped with modern facilities for dealing with road engineering problems. A knowledge of these resources should be of the greatest value to Military Commanders under conditions of emergency.

Through them they would have at their disposal a qualified and experienced organization to carry out such work as demolition, reconstruction of highways, communications, etc. Local Authorities also control in many cases such undertakings as water, gas and electricity, and are the responsible authorities for highways, public health, sanitation and public assistance.

(*c*) It will therefore be appreciated that in a socially highly developed country like Britain, where so much of the life of the country is organized under Local Authorities, any Military Officer who attempts to operate in an area without the closest co-operation with such Local Authorities will merely create chaos. Whereas, if he makes himself fully acquainted with the personalities and the machinery of the Civil organizations, he will find at his disposal facilities which will greatly assist him in any task which he may be ordered to carry out.

3. Work of the Civil Defence Services and of the Fighting Services in large-scale Air Raids

(*a*) INTRODUCTION

(i) In dealing with an air raid I always divide it into two parts: THE BATTLE—that is to say, the work that has to be done while the actual raid is in progress; and the AFTERMATH, which is the work that has to be done during the days which follow the attack.

I always speak of the BATTLE when referring to a big raid, and I am now talking of raids comparable to those I knew personally in Bristol and Plymouth, or to those at Coventry, Portsmouth, Southampton, etc., because I think it is necessary to impress on all concerned that a raid of this nature *is* a battle, and a very desperate battle too.

I experienced several battles of the first magnitude on the Western Front in the last war, and although these air-raid battles are not actually as dangerous, that is to say do not produce the same number of casualties, I believe they are just as nerve-racking. The falling bombs, the crashing buildings, the blazing streets, the noise of the A.A. barrage, and the fact that it is all taking place at night, make it a very terrifying performance.

So I call it the BATTLE in order that there may be no doubt in anybody's mind that when the Civil Defence Services are employed in it they are not merely putting out fires, attending to injured people, or looking after the homeless, but are taking part in a battle of the first magnitude.

Therefore, I also want to impress on everyone concerned that the organization, training, equipment and preparation must be considered to be just as important for the Civil Defence Services as for the Fighting Services.

(ii) In our fight against air raids we are principally concerned with—

(1) preventing damage as much as possible (damage by fire is said to be about ten to one of that by high explosive);
(2) repairing such damage as quickly as possible;
(3) and above all, keeping up the morale and the will to win of the ordinary citizen.

The last is particularly important if we are to prevent demoralization and to maintain our fighting spirit.

(iii) The fifth column, whether of the intentional or unintentional nature, does exist in this country and any measure of demoralization makes its work easier. It is no good thinking raids have no effect; they have, and our efforts must be constantly directed to counteracting that effect. It is said that Hitler is more interested in moral effect than material. That is not surprising, as Napoleon laid it down that the moral to the physical is as three to one.

(iv) Everyone is in this battle against air-attack, whether such attack is normal night bombing or a part of invasion—Civil Defence Services, Regular Army and Home Guard, any other services in the area, W.V.S., and other voluntary bodies, youth organizations, and in fact all except those who are debarred by age or infirmity. Only thus can we win this battle and save our cities from destruction.

It follows that the Services cannot take all the young and active man-power and expect the Civil Defence to fight this battle alone; first, because of the moral effect which I have explained to you and, secondly, because we have not sufficient efficient man-power left to fight it successfully alone. We have already been given definite military assistance for our docks as these are such an important feature in the Battle of the Atlantic. For it is obviously useless to preserve our ships at sea if the facilities for discharging their cargoes in this country are destroyed.

(*b*) THE WORK TO BE DONE IN THE BATTLE, AND IN THE AFTERMATH

(i) The first stage comes while the raid is in actual progress and consists mainly of fighting incendiaries, fighting fires, rescuing those buried by high explosive bombs, first aid, conveying the seriously wounded to hospital, taking the dehoused to hostels, supplying the workers with refreshments.

(ii) In the second stage the main work is care of the homeless; communal feeding; salvage of furniture and other belongings of those dehoused; billeting the homeless; public assistance, viz. food, money, clothes, pensions; searching for the missing; and answering enquiries both from people on the spot and by telegram, etc. An Information Bureau must be established and means of issuing instructions and propaganda arranged: loud-speaker vans, posters, bulletins in papers and so forth. I give priority to these things because they are so important for morale. Repair of public utility services, clearance of streets, demolition of unsafe buildings, removal of unexploded bombs, cordoning off of areas affected, must also have speedy attention.

(*c*) THE BATTLE

The air-raid battle is a defensive battle with, as in all defensive battles, a certain measure of counter-offensive action. Now, what is the army that fights these battles?

The forces engaged in fighting the enemy in this defensive battle consist of:

(i) The *Royal Observer Corps*, whose members plot the advance of the hostile planes, and who, in conjunction with certain other agencies, have brought information as to the movement of enemy planes to such a pitch of perfection. They even provide information while the raid is in progress.

(ii) The *Royal Air Force*, whose night fighters take counter-offensive action against the enemy planes.

(iii) The *Anti-Aircraft Batteries*.

(iv) The *Balloon Barrage*, which keeps the enemy planes at suitable heights for the A.A. guns and prevents low bombing.

(v) The *A.R.P. Control and Report Centres*, which receive and disseminate information from all sources.

(vi) The *Police and Wardens*, who, being in position and on patrol, give the first information and take first action.

(vii) The *Fire Guard*, which deals with incendiary bombs and incipient fires.

(viii) The *National Fire Service* which deals with the fires which have got beyond the capacity of the Fire Guards.

(ix) The *Rescue Parties, the First Aid Parties and the Ambulances* of the A.R.P. Services.

(x) Those who deal with the *care and welfare of the homeless* and man the Rest Centres.

(xi) The *Military*, consisting of those on the spot who work as organized bodies from their barracks or billets, or those at places of entertainment or in the streets, who should render immediate assistance to the Civil Defence Services; and such other reinforcements as the local Military Commander can throw into the battle at the request of the A.R.P. Controller.

(xii) The *Bomb Disposal Units*.

(*d*) An Army

Thus, we not only have to fight a battle, but to fight it with an army consisting of the various arms of both the Fighting and Civil Defence Services. If this battle is to be fought successfully, it must be fought by all those arms working in close co-operation from the moment the information is received that attacking planes are on their way until not only the last of them has left for home, but also the effects of their attack have been dealt with, the last fires put out, the last of the buried rescued, the last of the homeless cared for, and the true work of the aftermath begun.

Therefore, as we are dealing with an army, we must deal with it on military lines. That is to say, it must have a Commander and a Staff who can control and direct it and also control and direct such reinforcements as may be sent to it.

As I am now dealing only with the Civil Defence Services in detail, I shall confine myself to the Services listed as (v) to (x) above. But (xi) and (xii) which, though military units, are present entirely for the assistance of the Civil Defence Services, will have to be considered as part of them for the purpose of what follows.

(*e*) The Method of the Battle

The battle is fought on military lines. Information is sent in to the Control Centre of the town affected, and the Control Centre informs Regional Headquarters' War Room. First reinforcements are at the disposal of the local Controller on request at once. Major reinforcements are arranged by Regional Headquarters, who order them to stand to as soon as likelihood of their being required is seen. If reinforcements are wanted from outside the Region, Home Security makes the necessary arrangements. Normally, only reinforcements for the Fire Services are brought in from outside the Region. Reinforcements are brought to Reinforcement Depots on the edge of the battle and deployed as required. Reserves of personnel, especially for fire services, are often moved up from various parts of the country to nearby towns. They are to relieve tired men next day.

(*f*) Control and Reinforcement

(i) In a big raid no town can fight the battle with its own resources. It must have reinforcements. Therefore, close co-operation has to be established with the various authorities in the surrounding area on whose help it has immediate call, and with Regional Headquarters in the case of reinforcements which are to be sent from further afield. These reinforcements have to be brought to the battle and deployed in it. This is the crux of the problem; for the winning of this battle, as of any other, depends on such control and direction being maintained as will enable the troops in the battle to be used to the best advantage, the reinforcements being deployed when and where they are required, and those troops whose immediate task is finished being withdrawn, reorganized, refreshed and rested, so as to be ready for re-employment.

(ii) It must be admitted, I think, that in major air attacks, not only at the places which I saw for myself, but practically everywhere else, our control of traffic, our deployment of reinforcements and the control of the battle itself, have not been as efficient as they should have been. What struck me as a soldier, trained in both command and staff work, and especially in mechanized movement, was the way in which reinforcing units dribbled into the battle piece-meal, and probably reported at some place, A.R.P. Control or Depot, a Fire Station etc., usually right in the middle of the battle. The arrangements for their reception, welfare, instruction as to the situation, and deployment were usually not well organized. Consequently, they were frequently not used in the best manner or to the best purpose. In the Bulletin that I issued as a result of the first three heavy raids on Bristol, I said:

> 'No Military Commander would have his reinforcements sent into the middle of the battle; he would have them brought to some place in rear of the battle where they would be taken charge of by a competent commander who would receive instructions from Battle Headquarters as to how they were to be disposed. As with military reinforcements so with reinforcements of the Civil Defence Services. If they report for duty at a Control Centre in the middle of the area of operations, it is most difficult to deploy them as required. Therefore, reinforcements should report for duty at a rendezvous outside the area of operations and from there be deployed as required, the commander of that rendezvous being in touch with the Controller and receiving instructions from him as to the deployment of the reserves.'

This matter has now been remedied in most Regions by the establishment of Reinforcement Depots on the principal routes leading into the various target towns. Here all reinforcements will be received, refreshed, rested and prepared. Their arrival and strength will be reported to what I shall call Battle Headquarters and their deployment carried out as ordered from there. The details of this I deal with in (*g*) and (*h*) below.

(iii) As regards control of the battle, I saw on occasions, that the man in command, without a properly trained staff, was himself being inundated with messages which he was expected to interpret, form a picture from, and issue the necessary instructions on. This was an impossibility for anyone however competent and on these questions I wrote in the same Bulletin as follows:

> 'In the Army the Officer in command of operations always keeps away from the Command Centre where information is being received, in order that he may not receive details of information as they arrive, but a clear picture of events through his staff. Thus he receives the information which it is essential for him to have, and in a form which enables him to obtain a complete picture of events. No man in command who personally receives numbers of messages giving small details can possibly see the whole picture of the Battle as he ought to in order to enable him to make decisions and give orders. It is the same with the Chief Officer of any Civil Defence Service. He must be kept away from the hurly-burly of the message room and the messages must come to him filtered by a competent staff officer and be presented in such a manner that he, like the Military Commander, does not have to sort out a mass of detail, but is given a clear picture of events as they take place. Thus he, also, can keep his mind clear for appreciating the situation and issuing orders.'

(*g*) COMMAND

(i) The principles of military command and control which have been built up over centuries of trial and error lay down emphatically the principle of single control. This does not mean that each Commander of a separate service does not command his own service in the battle, but it does mean that the work of all these services, including their movement to the battle, their deployment to the battle and their employment in the battle, are all under one direction.

(ii) Therefore, it appears to me that we should establish a main battle headquarters and that at these headquarters there should be representatives of all the services concerned in the battle on the civil front. They should have a joint intelligence bureau which is linked up with the various points at which the intelligence is obtained and with the respective headquarters, both Fire and A.R.P.

(iii) Now that all reinforcements will be brought initially to reinforcement depots outside the town the work should be simpler, as the deployment will start from these depots. But there must be some control on the edge of the battle to receive these reinforcements and to deploy them in detail as required; for this control I visualize a sub-battle headquarters of the main battle headquarters suggested above. Therefore, this sub-battle headquarters also must be in touch with the reinforcement depots.

(iv) There may have to be two or more sub-battle headquarters because the reinforcements coming in to one side of the battle cannot be controlled by the sub-battle headquarters at the other side. Therefore, it is probable that such a headquarters will have to be established on each of the main routes into the town which will be used by reinforcements coming from the reinforcement depots.

(*h*) Traffic Control

Traffic control points must be established between the depots and the sub-battle headquarters so that, if streets become blocked, reinforcements can be diverted to other routes on their way in; this should obviate congestion and delay.

The traffic control for all reinforcements entering must be under the general direction of the police. It must ensure that the correct routes are used and that when the reinforcements reach the parking place selected near the sub-battle headquarters they are parked in such a manner that any service required can be brought up without delay and sent to its task.

When the reinforcements are actually deployed in the battle, their movements must be directed by the Commander of the service in question, but the traffic control must still keep general supervision as to parking, even in the battle, otherwise congestion will occur and their movements will be hampered. For example, I have seen Fire Service appliances (which are always the most numerous), ambulances, rescue lorries, canteens and other vehicles, jammed in narrow streets so that none could be quickly moved.

In this event, the operations of the Fire Service appliances are handicapped by congestion caused not only by their own vehicles but by the vehicles of other services. Therefore, control of the vehicles of all services by some over-riding authority seems necessary. This must be the Police in consultation with the Commander of the Service concerned.

(*i*) Communications and Intelligence

At present there is usually a considerable duplication of these Services, which results in far more messages being received at the various Control Centres than are necessary.

This is due to the various Services working too much in watertight compartments and failing to pool their resources and to have sufficiently good liaison between them. Therefore we must not duplicate channels of command, communication and intelligence. This is not to say that we must not prepare various means of carrying out these functions, but only that, as in the Army, they must be run jointly for the service of all.

(*j*) Incident Control

The general control of the battle includes Incident Control: that is to say there must be, under each sub-battle headquarters, a certain number of smaller controls which take charge of 'Incidents', i.e. of small areas of the battle.

The original idea on which incident control was based had in mind the possibility of isolated incidents, a bomb, or a stick of bombs, falling in some part of a town.

Now, however, in the case of large-scale air attacks, we have to deal with what may start as a number of separate incidents, but these incidents soon become joined up, or extended, and embrace large areas of the town attacked.

When this happens it is impossible to establish a large number of Incident Controls or Posts, and the minimum number of posts adequate to deal with the situation should be established in each area of the battle. These Posts should be grouped so as to work in communication with one sub-battle headquarters, a convenient number to each.

The incident controller, or incident officer as he is usually called, must have some definite place from which to send and receive intelligence and instructions.

Therefore the Incident Post is frequently set up at the nearest Wardens' Post. If the nearest Wardens' Post is not conveniently situated for control and speed of communications, another more suitable place may be selected. The place selected, however, must provide suitable accommodation and shelter for the incident officer, his assistants, messengers, etc., and telephone communication for as long as such remains in working order.

The National Fire Service has now been provided with Control Vans; these really constitute Mobile Incident Posts. Normally these vans and the Incident Posts should be in close proximity. Thus the services of both can be combined to the advantage of each, and a pooling of man-power, communications and other resources obtained. Without doubt each will be of far more value to the services as a whole than if used separately without close liaison between the two. And in this way we can secure a combined Incident Post at which all the Civil Defence Services are represented, and at which all pool their resources.

Thus we should have three forms of Control, main battle headquarters, sub-battle headquarters and incident control in the shape of Combined Incident Posts of Police, A.R.P., and the Mobile Vans of the National Fire Service.

At each of these centres of control, each service would be represented, and by this means unified direction, communication and intelligence should be obtained.

(*k*) Summary

The foregoing represents the general idea which, in my judgment, should be adhered to as far as the various factors and circumstances permit.

The main features of the idea are:

(i) Everybody belongs to one service for the purpose of the battle and works under one direction on one central plan.

(ii) There must be one main battle headquarters, at which all services are represented, which is supplied constantly with information by one intelligence system, which works for all.

This battle headquarters must be in touch with the A.R.P. and Fire Control Centres and with the Reinforcement Depots.

(iii) The requisite number of sub-battle headquarters, in control of an appropriate number of Incident Posts, is provided.

(iv) The greatest possible amount of pooling of Intelligence, Communications, Messengers and resources is secured.

(v) There is complete co-operation between all the Services engaged.

4. Military Assistance to the Civil Defence Services

(*a*) It has been found that in dealing with large-scale air attack the help of the Fighting Services is of the utmost importance. The importance of this help is increasing because the Civil Defence Services are becoming more and more depleted of their best men as these are called up for the Fighting Services.

We recognize that this call-up is essential and I personally believe that the larger the number of men trained in one of the three Services the better it will be for Britain, but if they are to be so called up and trained, the Services must reciprocate by helping the Civil Defence to the greatest extent in their power. We also recognize that, because the primary duty of both Regulars and Home Guard is repelling invasion, when invasion comes we shall have to fight the air raids which will accompany it by ourselves.

(*b*) The effect of assistance given by the troops to the Civil Defence Services is not only material, it has also a strong psychological aspect.

In the first place, the effect of disciplined and properly organized bodies of troops mixing with the Civil Defence Services during a raid, and assisting them in their tasks, has a great effect on civil morale. Our experience in Bristol and Plymouth showed that the presence of bodies of sailors and soldiers gives an example of courage, initiative and ordered procedure which stimulates and increases the same characteristics amongst the civilians. In addition, it is magnificent training for the troops. Operating as they are under the difficult and dangerous conditions brought about by falling bombs and blazing streets, they get a foretaste of battle conditions which is of high training value. Battalion Commanders have told me that their units have, from the standpoint of training for war, benefited enormously from such experiences.

In the immediate aftermath of a raid the sight of bodies of troops at work in the streets helping to clear up the mess, and to resuscitate the life of the community, raises the morale of the whole population to a degree which we know from experience to be of incalculable value.

(*c*) Military Assistance is of three kinds. *Immediate*: to fight the incendiaries at the start of and during the battle. The plan for this is arranged and rehearsed by the Military and the Civil Defence Authorities in conjunction. *Minor*: the Civil Authority calls on the local Military Commander during the battle for assistance, usually for fire-fighting, rescue, and assisting the police. This is arranged beforehand between the Civil Authority and the local Commander. *Major*: this is asked for by Regional Headquarters as a result of the first conference after the attack. It involves Royal Engineers, Royal Corps of Signals, pioneers, infantry, and traffic control personnel. The Military Commander for this is detailed in advance for each likely place. He works with the Local Authority and usually with the City Engineer.

5. Conference

(*a*) On the morning after an attack, a conference is held by the Emergency Committee of the town, at which will attend all appropriate officials of the town, representatives of the Civil Defence Services, Fighting Services, Post Office, Chairman of the reconstruction panel, and other interests concerned. A team from Regional Headquarters would also attend the conference and would normally consist of a Deputy Regional Commissioner, a Regional Officer for A.R.P., a Military Liaison Officer and Ministerial representatives, such as Health, Food, Post Office, Transport, Emergency Works, Information, and such other departments as may be deemed necessary.

(*b*) A review is made of the damage done, and the immediate requirements of the situation are met.

Sub-committees are formed to deal with the various matters and progress is reported to the Emergency Committee later in the day. Nature of work, labour and material required, where they are to come from, transport, housing and feeding of labour, military aid required, priority of work, etc., have to be decided.

(*c*) Later, a plan for the reconstruction of the damaged area having been decided upon, labour is drafted in from other places and material is provided, under arrangements made with the Central Authorities, so that the work of reconstruction can proceed.

6. Reinforcement Depots

Reinforcement Depots are places such as large houses and schools, selected outside the target area of a town, on, or near, a main road leading into such town, and reasonably close to it. They should have good parking accommodation for vehicles. Here arrangements are made for the reception, feeding and resting of reinforcements and for guides to take those reinforcements where required.

Each Depot will have a Commandant, who will have such staff, operational and administrative, as is required.

Each service will be represented on the staff of the Commandant.

The Depots will be in communication by telephone and despatch riders with the necessary control centres of the town.

All reinforcements sent to the town will report at one of these depots and, on completion of their work, will return to it for rest and refreshment, and subsequent re-employment or return to their normal station.

7. Effects of Bombing on Towns of various Sizes

In case of invasion many of the smaller towns, which have hitherto been immune, may well be subjected to severe attacks.

In computing the probable effects on any particular town of severe bombing attacks, the size and shape of such town must be taken into consideration. It may be taken as a rough guide that a high-altitude night-bombing attack by a couple of hundred planes will cover with its main weight, say 70 per cent of its bombs, an area about one mile long by half a mile wide, with the remainder of the bombs widely distributed outside that area.

If the larger part of the town is covered by such a yardstick, then it can be anticipated that the effect on the population will be very severe and that the number of people who will have to be catered for in the surrounding country will be a high percentage of the total of its inhabitants. If, however, the town is of such size and shape that only a small percentage of it is covered by such a yardstick, then the percentage to be catered for will be considerably less. If a town is surrounded by a large built-up area, the exodus is much less noticeable, and the situation far less acute, than if it is a small compact town surrounded by a rural countryside. A long narrow town will normally be badly affected; for, if such a yardstick be applied to it, it will be seen that the main part of the town is directly affected and the whole indirectly affected, not only by the bombs falling outside the main beaten zone, but by the fact that the whole town is so close to the beaten zone that the detonation of bombs falling in it can actually be felt by the inhabitants outside it. Thus the moral effect is greater than an attack of similar intensity upon a town which covers a larger area.

For this reason just as much care must be taken in preparation for the care and welfare of the inhabitants of the smaller rural towns as is taken for the larger cities.

8. Civilian Welfare Committees

(*a*) The heavy and continuous raids on Plymouth in April 1941, following on those in March of that year, led to an exodus from the town into the surrounding districts.

The human problem created by people seeking shelter outside a bombed city is a difficult one, and its solution requires the greatest goodwill, energy and organization. It is only by the Region, the target town and the other Local Authorities concerned pooling their brains, energies and resources that it can be solved satisfactorily.

When a town has been heavily attacked, all the Rest Centres in that town must be cleared the next day, and the occupants transferred to Rest Centres in the surrounding districts, or at any rate outside the target area.

Thus the City centres are made ready to receive a fresh inflow of people should a further attack occur, and being less crowded will not form such a dangerous target. If the town is attacked on several nights in succession, this clearance has to be done each day. Therefore the surrounding centres must be prepared for a further influx after each night's attack.

As a result of heavy and sustained attacks, we may be confronted by the problem of numbers of people trekking out into the country to seek refuge—of people who, although they may not come under the heading of genuinely homeless, have to be catered for in the first stage of the aftermath and certainly until confidence has been restored and until those people who are not homeless have been persuaded to return. It is no use to say that people in unofficial categories cannot expect to be looked after. We must look after them as far as it is possible to do so. However big the human problem may be, it must be faced and dealt with to the full extent of our resources. That is not merely a humanitarian matter, but a vital part of our war effort, the sustaining of morale and of the 'will to victory'.

(*b*) To meet this difficulty we formed in No. 7 Region what we called Civilian Welfare Committees. We began by forming them round Plymouth; then round other target towns, and now we have extended the organization to cover the whole Region. Thus we can cater for invasion conditions when smaller towns are attacked and rendered uninhabitable, as well as for normal night bombing.

In addition, the fact that we have prepared rest centres throughout the Region enables us to cope with almost any scale of attack and, in addition, to render help to people from neighbouring Regions if required to do so.

Meetings were held throughout the Region at which representatives of all Local Authorities, Fighting Services, Civil Defence Services, Police, Fire Service, voluntary bodies and special interests were present. At each meeting the object of these Committees was explained, and at each their formation was unanimously agreed to.

Except for Gloucestershire, which has a scheme of its own, our Counties are divided up into Zones of Local Authorities. Cornwall has two Zones, the other Counties have three each. There is one Civilian Welfare Committee to each Zone.

Each Civilian Welfare Committee consists of representatives of the County Council, of any County Borough in the Zone, of all local authorities in the Zone, and of the other bodies and interests mentioned above. These Committees deal with all matters affecting the housing, care and welfare of people affected by enemy action, and the appropriate representatives of Ministries and local transport facilities attend as required.

In this way it is hoped that every possible arrangement will be made to ensure that the least possible discomfort accrues to those who, in the future, may be driven from their normal habitation by enemy action.

(*c*) The organization of transport, by rail and road, to take the people from the target town to the rest centres must be put in hand.

The procedure is not the same everywhere, but that adopted by one large target town is as follows.

Each Zone Committee, through the medium of a small executive committee which remains in session for as long as required during the aftermath, keeps in touch with the target town and the local authorities concerned.

The target town keeps the Zone Committees informed of the numbers requiring accommodation, and the Zone Committee decides where they are to be sent. Thus the official evacuees can be sent to such places as will leave a cushion area near the town for the accommodation of any trekkers.

The official evacuees should be moved out by 2 p.m. or thereabouts.

I am also of the opinion, for the reasons I have given, that transport arrangements should be made for unofficial evacuees as far as is possible. If this is not done, great difficulty will be experienced in guiding these people to the places where accommodation is available, and overcrowding, confusion and unnecessary discomfort will result.

(*d*) While nothing in this life can be made perfect, these Committees are already bringing about the very results that we hoped for when we started the idea. That is to say, they have achieved closer co-operation between all the local authorities, voluntary bodies, the Region, the Ministries represented at Regional Headquarters, the Services, etc. In fact, reports from all quarters have been that these Committees are creating a unity of purpose and a collaboration to obtain the desired result that is producing not merely material results but a psychological atmosphere which I firmly believe will be of assistance not only in winning the war but also in what may be just as difficult a problem, the winning of the peace.

9. Women's Voluntary Services for Civil Defence

(*a*) The Women's Voluntary Services for Civil Defence were formed in June 1938, at the request of the Home Secretary, with the object of enrolling women for the A.R.P. Services of the local authorities.

Since that date over a million members, drawn from every creed, class and section of the community, have been enrolled and, at the request of the Home Office, the scope of the work has been enormously enlarged so that it now includes evacuation, clothing, transport, emergency work in Rest Centres, feeding, both the Civil population in British Restaurants and the Forces in canteens, driving the Queen's Messenger convoys, mobile canteen work, hospital supply services, salvage and many other things.

The pattern of the organization is that of the local authority; that is to say, there are an Administrator and an Organizer for each Civil Defence Region, County and County Borough, Organizers for the Counties and large Cities, and a Centre Organizer for each Urban and Rural District. In this Centre a Representative is appointed for every village, so that the link with the individual members may be complete.

A recent development has been the formation of what is known as the Housewives' Service. The object of this is to help the Wardens in time of stress, and the members, who are not required to work away from their own homes, undertake to do certain very simple things such as to know about the raid movements of their neighbours, always have a bucket of water ready, keep spare blankets in the house, etc. The service has proved its worth and is now operative in most large cities and many of the smaller towns and even villages. Its organization follows that of the Wardens' Service. There is a Divisional Organizer, and, under her, organizers for each group, post and street, and in every street there are two or more members according to its length. The Housewives do not have to enrol as full W.V.S. members, but before they can become eligible for a window card (the distinctive mark of a Housewife), they must have attended five lectures or talks on the following subjects: First Aid, Home Nursing, Anti-Gas, Local A.R.P. Organization and Emergency Cooking.

By means of the Housewives' Service opportunity for National Service has been given to many women who have never had it before, and also it has proved that when it is necessary to disseminate information or make a house to house canvass, this service, with its tentacles reaching down to the women in every street, is invaluable.

The position of the W.V.S. as a whole has perhaps been best summed up by Mr Herbert Morrison in a recent circular to local authorities in which he says that the W.V.S. is in the same relation to their Civil Defence Services as the Women's Auxiliary Services are to the Armed Forces of the Crown.

(*b*) I cannot emphasize too strongly the value of this organization to the Civil Defence Services and to the nation generally. It is not only the material assistance which they give; their contribution to the morale and cohesion of all sections of the population has been outstanding in every emergency with which we have been faced. The Women's Voluntary Services, together with the Housewives' Service, give us an organization which stretches from Regional Headquarters to the humblest dwelling in town or village.

I have emphasized the necessity for a United Front by all sections of the community—without it we cannot mobilize the whole of our mental and physical war effort. When the German military power collapses, Europe may well be in chaos and it is quite possible that for a while revolution, civil war and anarchy will prevail in many countries. Then the best hope for Europe will be a stable Britain using her influence, organizing power and resources to bring order out of that chaos. For this, as for the war, a completely united front will be essential, and it will be more difficult to achieve in peace than in war. Yet without it we shall fail in our task, Europe will collapse, and we shall collapse with it.

Every organization we build for the prosecution of the war must be developed on lines that will make it of value for the prosecution of the peace. We prepare to fight the battle and at the same time we prepare to deal with the aftermath. The same machinery, embracing a great variety of activities, prepares for both. In the same way, the same machinery that carries out the war must be prepared to carry out the aftermath of the war. To do this effectively, the greatest objective must be the maintenance of the United Front. There is no organization that will be of greater assistance in ensuring that all sections of the community maintain it than this new Civil Trinity of the Women's Voluntary Services, the Housewives' Service and the Wardens' Service working hand in hand. So let us encourage, and help, this great organization of women to develop its strength and influence to the fullest possible extent.

All other voluntary bodies, all organizations that deal with youth and all whose duty it is to educate and guide any section of the community must band together, plan together and work together to instruct all in the dangers and duties of the future. Only thus can we build a United Front both for the war and for the very turbulent peace that will follow it. Only thus can we save Europe from disintegration and only thus can we save ourselves from the same fate.

10. Morale

> The Master had been asked wherein lies the Strength of a State. The Master said, Sufficient food, sufficient weapons, and the confidence of the common people. Tzu-king said, Suppose you had no choice but to dispense with one of these three, which would you forgo? The Master said, Weapons. Tzu-king said, Suppose you were forced to dispense with one of the two that were left, which would you forgo? The Master said, Food. Danger of enemies, danger of starvation, such were recurrent perils which could be overcome if confidence were unshaken. But once let confidence go, neither arms nor food could save the country.
>
> Confucius.

(*a*) I have already stressed the fact that one of the most important functions of the Civil Defence Organization is to ensure that the ordinary population retains a high state of morale under all circumstances.

Victory or defeat in this war depends in the last resort upon the morale of the people, upon whether you and I, the ordinary citizens of this country, can hold to the 'will to victory' in the face of every misery, of every disappointment, of every loss, and of every discomfort.

If we, and our fellow-citizens, can do this, we shall win the war; if we cannot do this, we shall lose the war. If (to repeat our metaphor) the base of our triangle disintegrates, the apex will fall to the ground. In other words, if the Germans could break our morale, no strength of fighting forces, no supply of war material, could save us from defeat. That is why I say that the human element is more important than the material, and it is with the human element in particular that the Civil Defence Organization has to deal; for if it neglects that, then the rest of its work is of little value.

(*b*) My quotation from Confucius shows that The Master considered confidence the most important factor in the maintenance of morale. Confidence consists mainly, as far as the civil population is concerned, in three things.

Confidence in oneself, which really means confidence in one's ability to do one's job efficiently. Highly trained firemen and A.R.P. workers gain confidence in themselves from their very efficiency. The knowledge that you have a job to do and are competent to do it leads to the preservation of high morale. But confidence in oneself also comes from *confidence in one's leaders*, and *confidence in the efficiency of the organization* of which one is a part,

In a large-scale air attack there are a great many people who can have no job to do, and who perforce must become part of that permanently, or temporarily, homeless crowd who are in air-raid shelters, or rest centres. If the attack continues for several nights in succession, they may also form part of that crowd who are seeking safer quarters in the surrounding country. With these people the only tangible thing that helps to keep their morale from serious deterioration is confidence in the organization which is charged with their care and welfare. As long as it is obvious that everything possible is being done to help them, and that the organization for doing so is working efficiently, people will stand up to almost anything.

That is why I always maintain that the first and most important duty of the Civil Defence Organization is, as far as possible, to make perfect every arrangement for the care and welfare of the people both during the raid, and more important still, in the aftermath.

I have already pointed out the main things which have to be arranged for and I will not go into details of how these arrangements are put into effect. But there is one thing that requires long, widespread, and careful organization, and that is the care and welfare of those who leave the target town either because they are homeless or because for the time being they cannot face the possibility of continued attacks.

11. Conclusion

The foregoing shows generally the nature and variety of work which the Regional Organization has to carry out and supervise, and that no part of that work is of greater importance than ensuring that everything possible is done for the care and welfare of the people who live within its boundaries.

It is essential to understand that the Regional Organization aims at co-ordinating and assisting the existing forms of government. In fact, it aims at helping the government to do the work that has been imposed on it by the war and does not aim at doing that work for it. Therefore, the Regional Organization, whilst it has various powers which it can call into use when urgently required, is normally an agency for a system of co-operation on the spot and for the introduction of extra assistance from outside.

All departments of the Regional Organization must work together to eliminate the hampering effects of self-interest, vested interests, watertight compartments, and the secrecy complex. Their great objective must be the creation and maintenance of that United Front without which we can neither win the war nor preserve the peace. *Vis unita fortior*—strength, when united, is stronger.

Chapter V

THE UNITED FRONT FOR TOTAL WAR

A song for our Banner, the watchword recall
Which gave the Republic her Station:
'United we stand, Divided we fall',
It made and preserved us a Nation.

The Liberty Song, 1765, by John Dickinson of Philadelphia, 1732–1808.

An M.P. speaking at a Conference said:

Hitler has his invasion forces more or less intact across the Channel. Are we ready? Do the civilian population know what to do in the event of an attempted air-borne invasion in the spring? Do they know individually if they have to follow the 'guerrilla', the 'scorched earth' or the 'stay put' policy?

Our army gets its orders every day. It is a long time since the civilian army had theirs. *Bristol Evening Post*, 3 January 1942.

There must be no more passing of files....

Every officer must accept full responsibility in taking decisions....

The essential thing is speed in action.

Officers who show they cannot take responsibility should be replaced by those who can.

Sir Shenton Thomas, Governor of the Straits Settlements, January 1942.

1. Introduction

In my first lecture I tried to show you the features of modern war, that it is world-wide, mechanical, totalitarian, and Blitzkrieg. The result of these features is that everybody and everything is involved in it, that nobody and nothing escapes its effects. No longer are there two separate fronts, a civil and a military; now they must be as one, for each is dependent on the other. To win they must be one united front for total war. Speed is vital, not only in the air, on the ground, and on the sea, but in the office, in the factory, in the Council Chamber and, above all, in the mind.

Then I dealt with the evolution of modern war: how, through an insufficient or incorrect study of history, we failed to appreciate what that evolution would be, though in the pages of history it was written clearly enough for those who had the background of study and experience to interpret their reading correctly. This was combined with a form of pacifism which not only refused to prepare the country even to defend itself against attack, but tried to put the world right by leagues and words and formulae, and which had nothing behind it but a weakness which could enforce nothing, defend nothing and be nothing, except a source of trouble and danger to ourselves.

A study of armoured war shows us that whether such war is offensive or defensive, the same characteristics are required. These are a perfect cohesion, a perfect understanding of the plan by all from the general officer to the private soldier, a mutual understanding by all of each other's part in the plan, and yet an organization so flexible and so capable of decentralization that rigidity is abolished and full scope given for individual initiative, enterprise, decision and action. That is the secret of the successful conduct of modern war, and it is based on the highly trained individual who is so well instructed regarding the general plan, and his part in it, that he can be relied upon to react correctly to the ever-varying situations and ever-changing factors of the modern mechanized battle.

This is the only system of command that can implement the Blitzkrieg, and it is the only system that can defeat the Blitzkrieg. It is an attitude of mind on the part of both commanders and commanded and it means that the former take the latter fully into their confidence, trust the *man on the spot*, and forgive him anything except inertia. Thus the best is got out of every individual.

On the civil front it is the same. Here also failure to apply these methods will lead to uncertainty, hesitation and failure in emergency. So let us examine our organization on the civil front and see how far it is applicable to total war, and if it is not fully applicable, what has to be done to make it so.

2. Unity of Organization and Direction

As the text to my first lecture I took a quotation from General von der Goltz, who wrote: 'The principles of war are eternal, but the factors with which they have to deal are undergoing an incessant evolution.' Under our organization for fighting the war, whether it be our military organization or our civil, we have (as I said) to remember that, while the principles may be eternal, the factors with which we have to deal are all the time undergoing an incessant evolution, an incessant change. Therefore if we are to meet them we must organize ourselves, whether it be for civil defence or civil government, in a flexible manner that can adapt itself to, and meet, the ever-changing circumstances.

My experience of the situation caused by the successive aerial attacks on Plymouth showed me that the normal system, by which the varied affairs and interests of the community are the concern of many separate authorities, with little machinery by which unity of organization and direction can be achieved, is not fully effective in an emergency of this nature.

I compared this in my mind with the military organization which is the growth of over two thousand years of trial and error by the great captains. In the early days armies were merely collections of small tribal parties under their local chiefs who came together for a battle and, having collected the loot, dispersed again after it. From the time of Cyrus the Great, however, the military organization has been gradually evolving and it is designed to enable the supreme commander to direct the operations of large bodies of men for a specific purpose. Therefore that commander must work through a definite channel and definite hierarchy of command, and that channel and the organization which the commanders have to direct must be of a flexible nature and so organized that the will of the commander can be conveyed to the smallest unit concerned. Yet that smallest unit must be so organized and commanded that it can work on its own initiative in carrying out the plans of the higher commander. This is achieved by the fact that no higher commander is given more than a few subordinate commanders to deal with directly and no subordinate commander is given more than a few units to command, and he only commands them through their own commanders. In our civil organization we have a large number of autonomous and semi-autonomous units in the shape of local authorities with whom we attempt to deal through one higher authority. Therefore control is made difficult and the organization is not sufficiently flexible to meet the changing circumstances.

I can best illustrate what I mean by quoting from a letter to *The Times* in which, among other things, it was stated that the striking difference between the organization of the Army and that of certain civil organizations was that in the former decentralization is effected by the delegation of authority in descending scale through the divisional, brigade, battalion and company commanders, thus ensuring the full use of the ability of every person forming a link in the chain. It went on to say that sound administration means the maximum of supervision with the minimum of interference, and that the war will only be won by giving full scope to experienced administrators down to those of the smallest units concerned. We must ensure that, in the words of the letter, 'full use is made of the ability of every person forming a link in the chain' and that 'full scope is given to administrators' down to the smaller local authorities and even parishes.

It might be argued that the ideal form of government for waging a war of the first magnitude is a totalitarian one. It may be that under such a government the co-ordination of the efforts of a whole people is easier than it is under a democratic form of government. I return to this shortly, and I will only point out here that we have been given by the Services, in the North African campaigns of Generals Wavell and Auchinleck, splendid examples of democratic co-operation. In these campaigns there were three separate commanders-in-chief, and three separate services. As always a trinity. It is a curious thing that nearly everything in life resolves itself into a trinity of some kind. Yet in these campaigns, as in the words of the Creed, there were not three commanders-in-chief or three services, but one commander-in-chief and one service. One United Front for Total War. They planned together, trained together and fought as one. From this we ought to draw a great lesson in co-operation. Let us apply it to our civil organization.

3. Civil Organization

The local government of this country is truly democratic. It has been built up gradually over a period of more than a thousand years; from the time of the Saxons to the present day it has been constantly evolving. Like most things that have evolved in this way it is very much of a patchwork, and it is often difficult to understand where the responsibility of one authority begins and that of another ends; where one authority is entirely responsible and where it shares its responsibility with another. I suppose that in some ways it might be considered to be an unsuitable organization for waging a war of the first magnitude. On the other hand, I believe in it, because although it may not be the perfect machine, as perfect a machine for the great national effort as a totalitarian system may be, it has this enormous advantage, that it has trained thousands of people in the country to be able to 'run their own show', and, therefore, has produced thousands of people who are capable of being leaders and of developing that policy of improvisation in the face of changing circumstances for which, as a nation, we have always been famous. I am sure that this is due to the form of government that has grown up all over this country in the centuries that have passed. But it has the inherent defect of being composed of a number of entities, or semi-entities, and it can only be made to work if all these entities will co-operate and co-ordinate their efforts so as to achieve the unity of command and organization which is deemed essential in the fighting services.

As I have shown you from my brief military analogy, some form of re-grouping seems necessary, so that the higher authority will have fewer commanders and fewer units to deal with directly. Thus will the organization as a whole be more easy to direct and more flexible, and thus will individual units be able to work together on a general plan with more initiative and more executive and creative freedom. In recent times there have been articles and letters in *The Times* and other papers, and speeches in Parliament, most of which set out to prove that it is impossible to conduct a modern war with our present system of local government. Many of the writers suggest that only by placing an over-riding authority in charge can we deal with the problems created by such a war as the one we are now waging. That our democratic system of local government is being put to a very severe test I admit, but I think that if it is given certain further assistance for collaboration it will prove itself successful under the present conditions. I have already said that I am a strong believer in our general principles of democratic government; surely its survival is one of the things for which we are fighting. But it probably is not, in its present form, entirely suitable for waging total war owing to the difficulty of getting full liaison and co-operation between all the various bodies concerned, which is mainly due to the fact that there is no common body on which they all meet and no common channel of control and direction.

It is these deficiencies which I suggest can be overcome by the system of Civilian Welfare Committees already described, and by the system of Defence Committees which I am about to describe. Given these small extra organizations I think our system can be made suitable for waging total war, and can prove itself in the long run a match for the totalitarian system. This extra means of collaboration will strengthen and not weaken the work of local authorities, and give more and not less responsibility to them. Therefore I consider that, although my suggestions do advocate a certain measure of control and direction, they will produce a no less democratic system, and one giving greater mutual collaboration and assistance than we have at present.

I will quote from the speech which the Prime Minister made in London on 14 July 1941:

> 'How complex, sensitive, resilient, is the Society we have evolved over centuries, and how capable of withstanding the most unexpected strain.'

4. Defence Committees

It had always been my idea that when the Civilian Welfare Committees had got into working order a further step in collaboration with other bodies should be taken. The trouble has always been that we are organized vertically in each service and not to a sufficient degree laterally between each service. That is to say, we are inclined to work on parallel lines and in water-tight compartments. Parallel lines never meet and working in water-tight compartments, whether in military or civilian matters, can result in nothing but failure. In many ways our organization is like a wheel which has no rim. There often seems to be little connection between the spokes except at the hub. In consequence, instead of collaboration at each level between all concerned, where the situation can be mutually discussed, decisions taken and instructions given, we seem to have an over-centralization at the hub. This leads to constant reference back, and to lack of initiative and action at the lower levels. Thus we get delay, uncertainty and failure to take action.

This is exactly the opposite to the German doctrine of total war. They recognize that a reasonable plan executed with speed and decision is better than a better plan which, through delay in its implementation, may well be put into operation too late to be effective in situations caused by the speed of modern war. It is the fault not of individuals but of the system. This system has the effect of curbing self-reliance and initiative. It has a mental effect which is detrimental to the war effort. It makes people unwilling to act unless they get a direct order or are certain that they are acting within the four corners of some regulation.

The system which produces this attitude of mind also produces that habit of working in water-tight compartments, of over-emphasis on secrecy when most of the so-called secrets should be known to all if understanding and co-operation are to be made possible, and of failure to collaborate at all levels from the top to the bottom of our organization. It is this collaboration at every level that we must strive for. We have seen that the defence against the modern attack is the Web Defence. We must therefore, in our civil organization, substitute the web for the rimless wheel. We must organize the inhabitants of Britain to defend our fortress aerodrome, and the civilian and military effort must be co-ordinated for its defence.

The results achieved by the Germans are due to the intimate co-operation and understanding that exist between every branch of their fighting forces. When it comes to the defence of this island, from whatever kind of attack it may be, there can be only one combined defence to meet it, because the civilian and military effort will be so interlocked and interdependent that there must be the closest and most intimate collaboration between them. Only by complete mutual association and understanding before the emergency arises can this be produced. Therefore, when the military authorities asked the Regional Commissioner to organize Defence Committees in every civilian unit, small and great, throughout the Region, I welcomed the idea as I saw in it the best way to achieve the end which I had had in view, namely, the most intimate collaboration and co-operation between every branch of our forces, civilian and military.

With this object in view, Defence Committees were set up in the South-Western Region by every County Council, County Borough, Borough, Urban and Rural District; and they were formed also in every Parish or amalgamation of Parishes. Thus a comprehensive organization of these Committees was established throughout that Region.

As a link between the Rural Districts and their Parishes, a meeting of the Chairmen (and other members as desired) of Parish Committees is summoned from time to time so as to define their duties, and give them information, direction and advice. Such meetings ensure that the members become well acquainted with each other and obtain a good working knowledge of the various Services and the part each Service will play in conjunction with the other.

If a scheme of this nature is carried out in the right spirit, it should produce the unity of purpose and co-operation that we desire. In addition it will form a chain of organization, information and direction from Region to Parish. It will also, through its military members, produce a similar chain from the Higher Military Command to the Parish. This organization should also ensure that there is a plan known to all, that everyone knows his duties, and that leadership is ensured.

Note. Normally the following should be represented on a Defence Committee: Local Authority, Fighting Services, Civil Defence Services, Police, Fire, Women's Voluntary Service and other appropriate Voluntary Bodies, Special Interests.

5. Some Aspects of the Work that has to be done

(*a*) Owing to the variations in conditions and factors throughout the country, it is improbable that the same machinery will be applicable everywhere. Therefore the system of Civilian Welfare Committees and Defence Committees which I have described, and which works well in the South-Western Region, may not be suitable elsewhere. The work has to be done, but the method of doing it must be left to the man on the spot, for only he can know the local conditions and factors. So I will now merely state what I consider to be some of the most important things to be done if we are to prepare the inhabitants of Britain for the grave emergency that would be produced by invasion, and to create that united front without which victory cannot be achieved.

(*b*) We must prepare the people mentally as well as materially, for if the mental attitude towards anything is wrong, no material assets will bring success. In the military world we always say that surprise is one of the most potent factors for victory. But surprise is not merely a material thing. It also consists in being mentally surprised by the enemy using any new weapon or new technique; tanks, dive-bombing, gas, were all in their day a surprise. Now, the main object of this kind of surprise is not so much its material as its mental effect. Very often the first effect of such surprise is panic even among trained and disciplined troops. Dive-bombing had this effect on certain units, but those same units, once they had been taught how to conduct themselves and how to counter dive-bombing, soon took it quite philosophically and showed no signs of panic. Therefore the great thing for morale is to accustom the men to expect the danger and to prepare the men to give the proper reaction to that danger. One of the worst forms of training for morale is to minimize the danger, or to allow people not to expect it. In either of these cases the enemy attains surprise, that is to say, the recipients of the attack are mentally surprised by what happens. This is because they have not been prepared beforehand to expect it. Therefore they must be so prepared.

In the same way the ordinary population must be prepared. They must not be surprised. The civilian for his part must teach the military the ramifications of civil defence. In every military exercise I have seen there has been a lack of knowledge of this subject which would militate against success, deprive the military of much help that is really available and cause a lack of co-operation between the military and civilian sides of the defence force. For these must act as one force, not two, if the full strength of the country is to be available in an emergency. We must see that all lack of mutual knowledge is a thing of the past. We must break down that over-emphasis on secrecy which prevents us all knowing what the others are doing or going to do and thus is a bar to that mutual co-operation which is so essential.

Similarly the military must teach the civil population the nature of modern war, what to expect in case of invasion, how to conduct themselves so as to help and not hinder military operations, how to be possessed of high morale themselves and how to imbue others with the same attribute. The military will have great responsibility in this matter. We must remember that we are all in this war, the civilian just as much as the soldier, the women as the men, the children as much as either. The material element is of great importance in modern war, but the human element is still the vital factor. The human element is composed of spirit, mind and body, and while it is important that the body should be fit and that the mind should be alert, knowledgeable and courageous, it is the spirit that pervades and transcends all else. The doing is important, but the spirit of that doing is vital. The body may tire, the mind may know fear, but if the spirit is strong, the will to victory will remain. Therefore we have no greater task than to see that all with whom we come in contact are imbued with the right spirit.

6. Decentralization

How can we attain the ideal of making full use of the ability of every person forming a link in the chain of our organization, and how can we give full scope to administrators down to those of the smaller Local Authorities and even Parishes? How can we stimulate initiative, decision, and speed of thought and action?

It seems to me that the only way to achieve this end is by decentralization. I believe that our civil machine can never work with full efficiency in a crisis unless every local authority, however small, is taken fully into the picture, given a definite part to play, and 'under the maximum of supervision' but with 'the minimum of interference' allowed to play this part with initiative and vigour. It is essential, indeed, that all concerned should be working on one definite plan laid down by higher authority and that all should have the same objective in view. But in a crisis it is no less essential that the lower authorities should be allowed to carry out their part in the plan in such a manner as to fit in with the local circumstances and conditions. To carry out any operation of war we must have a certain degree of regimentation. It must, however, be the regimentation of the plan and of the machine that carries it out, but not of the individual. While all must thoroughly understand the plan, and the part which he himself and the others with him play in that plan, there must always be such decentralization as will allow the man on the spot to act with judgment and initiative. Too great regimentation is fatal to initiative. This is illustrated by an extract from an article in a Military Journal on the subject of 'the Italian Failure'. It says:

> 'Regimentation such as that to which Mussolini subjected the Italian people, though of considerable value in organization and control, is detrimental to character. It does afford the leader higher material strength and a greater ease of manipulation, but it blunts the weapon and deadens the spirit.'

The Germans in the last war were very prone to over-centralization and to treat the individual rather as an automaton than as a person capable of exercising judgment and individuality. Their training now is entirely different, as is shown by the following extract from an article in the Journal of the Royal United Services

Institution on the Tactics of the German Army, which I have already quoted in part (p. 21 above):

> 'We have already discussed the way in which Germany decentralizes the control of supporting arms and pays great attention to the training of the individual down to the junior N.C.O. and the private soldier. As might be expected under such circumstances, command and control are also decentralized on the principle that the man on the spot can far more easily judge what action should be taken, and must be prepared to exercise his judgment and use his initiative. Very great stress is laid on the need for initiative by even the most junior commanders. Truppenfuehrung—the equivalent of our Field Service Regulations—states "mistakes and even disobedience are better than inertia", and again "better a faulty plan which shows boldness and decision than a perfect plan enmeshed in uncertainty".'

Just as striking is the group initiative shown by the Russians, both in their defensive and offensive operations.

So we must ensure that we have a plan which everybody understands, and such collaboration between all concerned as will enable that plan to be put into effect, whatever the crisis may be, with initiative and judgment by the man on the spot.

7. Conclusion

As a nation we are not yet fighting in this war. Our troops in Malaya and in Libya, our Royal Navy and Merchant Navy and the R.A.F. are fighting. They are often fighting at a serious disadvantage. Why? Because the British Home Front is not fighting. It is not organised for Total War; it is not activated for Total War. It is engaged in limited war. In this respect the Russians are setting us an example. In Russia the home front is disciplined and organized in a manner which we accept in our fighting services, but which is apparently incomprehensible to our civilian population, or so our politicians believe.

National News-Letter, No. 289, 22 January 1942.

Sir Stafford Cripps commented on the morale of the Russian people as being the greatest single factor in the Russian triumph. 'The key of Russia's success is the whole-hearted loyalty and determination of every individual person in the land to defend his country.'

The Observer, 25 January 1942.

What it seems impossible to get the people of this country to understand, even now, is that this is total war in which every one of us, civil and military, man and woman, has to play his part. The invasion of Britain will be the acme of total war, and if the civil and military fronts are not united and co-ordinated into one total front we shall fail. If the civil front fails the military front will fail also, for the military will fight on the battle-field of the normal administration and life of the country. Their battle-field is the habitations of the people.

I have made a study of modern mechanized war. The Blitzkrieg is a thing of many parts, political and economic, fifth column and rumour, uncertainty and terror, planes and tanks, and all the rest of the modern mechanized technique. The great fundamental of its success is the breaking down of the will to win of the ordinary citizen. If we can prevent it doing that we have deprived it of one of its essential parts.

Now there are two things which are particularly likely to lead to such a break-down—ignorance of what is likely to happen and of what it is proposed to do to meet that happening, and lack of a definite job to do when it does happen. So we must ensure that people are not ignorant, that the civilian and the soldier fully understand the part that each will play, and that all feel that they are one army, fighting one battle, and fighting it for one purpose—the destruction of the enemy wherever he may be. In fact one United Front for Total War.

Transcending all is the mental, the psychological, aspect, which is what will stop panic and give to all a feeling of aggressive determination to win at all costs. For this reason the more people, and the more interests, that are drawn in the better. Not only officials but the ordinary non-official citizen must play his part in all preparations. For then mistakes are disclosed as they seldom are by officials alone. We must together plan, prepare and educate. The fact that the Committees, Conferences, or whatever you like to call them, which are summoned to achieve this object may be large is a good thing, for it means that more people are made to feel that they are playing a vital part in the war effort; it means that more active assistance is obtained, and that knowledge is more widely disseminated. Sub-committees carry out the actual work and a small executive committee plans the scheme and functions in emergency.

When the invasion comes, it will, at any rate at first, be principally air-borne. So all Britain will be a battle-field; towns large or small, aerodromes, factories, etc., will become vital fortresses, containing not only static defenders but the combat teams and guerrilla bands who must issue forth to harry and harass the enemy at all times, at all places in the vicinity, and by day and by night. Only thus can we make possible the work of the counter-attack troops, the counter-strokes of the field force. This is the essence of the defence against the Blitzkrieg; a network of islands of resistance throughout the area attacked. Thus in every one of these islands, large or small, a representative body of all sections of the community must meet, so that they, all together, plan the total defence in peace, and they, all together, carry out the total defence in war.

Thereby we arrive at an understanding of everyone's part of what total war means, of what they may expect, what the total plan is, and how it is to be put into effect. This will engender a war mentality, a confident and aggressive spirit, and a general acceleration of the whole war effort.

Our civil organization has many virtues, but it also has its defects, and some of these are centralization, which means that by not trusting the man on the spot you deny him scope for individual initiative, enterprise, decision and action: vested interests, which make people more concerned with the interests of their own particular show than with collaboration with others: an over-emphasis on secrecy, which leads to departments, services, etc., becoming what I call secret societies, so that co-operation is well-nigh impossible: and no chain of direction. This last is necessary, for, whether you wish to issue instructions or information or to educate, you must have a channel, and one recognized channel, from top to bottom through which to act.

That is why, in the South-Western Region, we established our Committees at every level of local authority from region to parish.

The idea behind our effort must not merely be to produce certain material results in an emergency, but to act on the minds of the people as well as their bodies, produce an atmosphere of war understanding and war effort, and a feeling that everyone has his part to play in the war. People must be made to realize that it is of the utmost importance for the prosecution of the war, and in preparation for the peace, that the most intense spirit of co-operation should prevail, and an atmosphere of mutual assistance, understanding, and collaboration be created. If all are imbued with this spirit we need have no fear that we shall not overcome our present difficulties or that those that will certainly arise when the war is over will not be satisfactorily settled.

Thus the public will be educated, the danger of ignorance mitigated, and by finding tasks for all, a feeling of confidence and determination will be engendered. Thus a spirit of co-operation and mutual trust will be created and the man on the spot, knowing that he is trusted, will strive to make himself worthy of that trust. Thus self-interest, vested interests and secrecy will be replaced by the feeling that nothing is worth while except to work for victory. Thus, and thus only, can we create that United Front that alone can ensure a total victory in the war, and after the war that strong and united Britain which can use its strength, influence, and resources, in helping to create not only a better Britain but a better world.

> '*Men pass away, but people abide. See that ye hold fast the heritage we leave you. Yea, and teach your children its value, that never in the coming centuries their hearts may fail them or their hands grow weak. Hitherto we have been too much afraid. Henceforth we will fear only God.*'
>
> Thus Francis Drake to Queen Elizabeth.

For EU product safety concerns, contact us at Calle de José Abascal, 56–1°, 28003 Madrid, Spain or eugpsr@cambridge.org.

www.ingramcontent.com/pod-product-compliance
Lightning Source LLC
LaVergne TN
LVHW090417210926
848480LV00002B/592

* 9 7 8 1 1 0 7 4 1 8 8 4 4 *